Engineering Applications of Computational Methods

Volume 5

Series Editors

The book series Engineering Applications of Computational Methods addresses the numerous applications of mathematical theory and latest computational or numerical methods in various fields of engineering. It emphasizes the practical application of these methods, with possible aspects in programming. New and developing computational methods using big data, machine learning and AI are discussed in this book series, and could be applied to engineering fields, such as manufacturing, industrial engineering, control engineering, civil engineering, energy engineering and material engineering.

The book series Engineering Applications of Computational Methods aims to introduce important computational methods adopted in different engineering projects to researchers and engineers. The individual book volumes in the series are thematic. The goal of each volume is to give readers a comprehensive overview of how the computational methods in a certain engineering area can be used. As a collection, the series provides valuable resources to a wide audience in academia, the engineering research community, industry and anyone else who are looking to expand their knowledge of computational methods.

More information about this series at http://www.springer.com/series/16380

Linmi Tao · Atif Mughees

Deep Learning for Hyperspectral Image Analysis and Classification

Linmi Tao
Department of Computer Science
and Technology
Tsinghua University
Beijing, China

Atif Mughees
Department of Computer Science
and Technology
Tsinghua University
Beijing, China

ISSN 2662-3366 ISSN 2662-3374 (electronic)
Engineering Applications of Computational Methods
ISBN 978-981-33-4422-8 ISBN 978-981-33-4420-4 (eBook)
https://doi.org/10.1007/978-981-33-4420-4

This Springer imprint is published by the registered company Springer Nature Singapore Pte Ltd.
The registered company address is: 152 Beach Road, #21-01/04 Gateway East, Singapore 189721, Singapore

Preface

With the rapid development in the field of science and technology, the Hyperspectral Image (HSI) analysis, being extensively broader and advanced technology, has acquired the wide as well as significant advancement in the conceptual, theoretical, and application level and has become an established discipline. It takes into account hundreds of contiguous spectral channels to discover earth resources that typical traditional vision sensors are unable to identify.

This book is an outcome of research efforts in machine learning based HSI processing over a decade, which is a new methodology from the HSI perspective. The title of the book "Deep Learning for Hyperspectral Image Understanding" represents that the main purpose of the book is to explore and define novel machine learning methods and techniques for the analysis and classification of the hyperspectral remote sensing scenes by incorporating spectral and spatial characteristics of the image. In particular, the prime target is on investigation and optimization of deep learning based deep feature extraction strategies. Furthermore, in the last chapter, we present the unsupervised traditional technique of sparse coding to effectively extract spatial features and design a framework to detect the redundancy and noise in the high-dimensional data. We hope the readers can experience both the merits and demerits of supervised deep learning versus the unsupervised sparse scheme in HSI area.

An important factor that makes this book different from other HSI books is that this book exploits theory, application, and analysis of HSI, starting from noise detection removal, deep learning based feature extraction and finally classification. The book is structured according to the deep learning methods in a sequentially associated chapters that are logically related to each other and can be studied backward and forward for additional details. More specifically, most of the experiments, simulated results, graphs, and experimental analysis comparisons have been organized to have a consistent and logical organization of both the HSI content and learning methods. Techniques are combined in such an integrated structure that readers can easily understand how the concepts were established and evolved.

This book can be considered as a complete recipe that covers techniques for HSI analysis. Some of these techniques such as unsupervised HSI noise detection removal, segmentation, and feature extraction are established and matured for practical implementation. They are evaluated and analyzed with extreme detail. Various deep learning techniques established in the book will also become really useful for the coming years. For this reason, we have made the book self-sufficient so that readers can effortlessly understand and implement the algorithms without much struggle. In doing so, we have incorporated comprehensive mathematical sources and experiments for explanation.

Tsinghua, Beijing, China Linmi Tao
August 2020 Atif Mughees

Acknowledgements

We owe much recognition to people who deserve our heartfelt appreciation. These individuals are my former Ph.D. students, Dr. Sami ul Haq, Mr. Xiaoqi Chen, and Dr. Rucheng Du. This book cannot be concluded and completed without their efforts and contributions. We would like to deeply thank Dr. Sami ul Haq for his valuable Ph.D. research on sparse coding, which is presented in the book.

This book comprises HSI work that has been researched and completed over a decade in the Department of Computer Science and Technology, Tsinghua University; BNRist; and Key Laboratory of Pervasive Computing, Ministry of Education; Beijing, China.

Finally, we thank the National Natural Science Foundation of China for the fundings under Grant 61672017 and 61272232.

Contents

Chapter 1
Introduction

The human curiosity to discover and apprehend the universe always results in expanding the limits of science and technology, remote sensing is yet another addition. Remote Sensing (RS) is the area of science that deals with observation, collection, and analysis of information associated with objects or events under study, without making physical contact. The launch of the first satellite in 1957, opened new doors for a wealth of information particularly for Earth Observation (EO). Space-borne and airborne platforms equipped with powerful sensors make it possible to acquire detailed information from the surface of the earth. Hyperspectral imaging sensors have the capability of capturing the detailed spectral characteristics of the received light in the sensor's covered area.

Image spectroscopy also known as hyperspectral imaging is a process of measuring the spectral signatures/chemical composition of the scene under airborne/space-borne sensor's field of view. Hyperspectral image comprises of detailed spectral and spatial information of each material in a specified scene. Sensors capture hundreds of narrow, contiguous spectral channels in the wavelength range of visible through near infrared hence provide huge spectral and spatial information of the surface of earth. Each pixel comprises a vector, where each value corresponds to a particular spectral signature across a sequence of continuous, narrow spectral bands and also contains detailed spatial characteristics. The size of the vector is equal to the total number of spectral channels that a particular sensor is capable of capturing. In case of hyperspectral imaging sensors, they are capable of acquiring hundreds of spectral channels. The availability of such a detailed information makes it possible to accurately discriminate materials of interest with enhanced classification accuracy. Classification of high-dimensional hyperspectral data is a challenging task.

L. Tao and A. Mughees, *Deep Learning for Hyperspectral Image Analysis and Classification*, Engineering Applications of Computational Methods 5,
https://doi.org/10.1007/978-981-33-4420-4_1

1.1 Applications of Hyperspectral Images

HSI have been widely used in analyzing the earth's surface due to its high distinctive capability to classify and discriminate different materials, which in turn has opened new doors for a vast range of applications such as mineral detection, precision farming, urban planning, environmental monitoring and management, and surveillance. In the past decade, hyperspectral imaging has pushed the science boundaries by providing a great deal of information and solving challenging problems such as scene analysis [2], environmental changes [3], and object classification [4]. Hyperspectral Imaging contains numerous applications. The power of full spectral information combined with the rich spatial information opens up enormous capabilities such as

- *Agriculture—crop identification, area determination, and condition monitoring:* HSI consisting of fields, crops can be employed for precision agriculture to manage and monitor the farming process. The hyperspectral images can be utilized for farm optimization and spatially enabled organization of procedural operations. The data can assist to locate the area and level of crop stress and then can be utilized to optimize the use of agricultural chemicals. The major application involves crop classification, crop damage assessment, and crop production estimation.
- *Ecological Science:* Hyperspectral images can be used in the classification of distinct ecological regions, based on their geology, structure, soils, plants, environment conditions, and aquatic resources.
- *Geological science:* Hyperspectral image techniques are now being increasingly utilized for preparing geological maps and extract the basic geological material which is utilized for further detailed analysis.
- *Surveillance and Military Applications:* Enriched spectral-spatial information contained in hyperspectral images has enormous military and surveillance applications. It enables us to keep a watch on military build-ups, and troop movements in the area under study.

Remote sensing scene can also be seen as a pile of scenes taken in diverse wavelengths (spectral bands), which results in a hyperspectral image. Specifically, each spectral channel of the hyperspectral data cube represents a gray-level image as presented in Fig. 1.1. The size of the 3D data cube is $L_1 \times L_2 \times S$ where $L_1 \times L2$ is the size of each spectral channel while S is the total count of spectral bands. More specifically, hyperspectral scene is divided into spectral and spatial characteristics.

1.2 Challenges in Hyperspectral Image Classification

Classifying HSI image into its land cover classes is fundamental, crucial, and decisive advantage of hyperspectral scene and has many major applications in almost all fields. However, supervised classification techniques in general, require plenty of labeled samples for different stages such as training, testing, validation, and parameter fine-tuning. This problem becomes more serious in remote sensing as it is very difficult

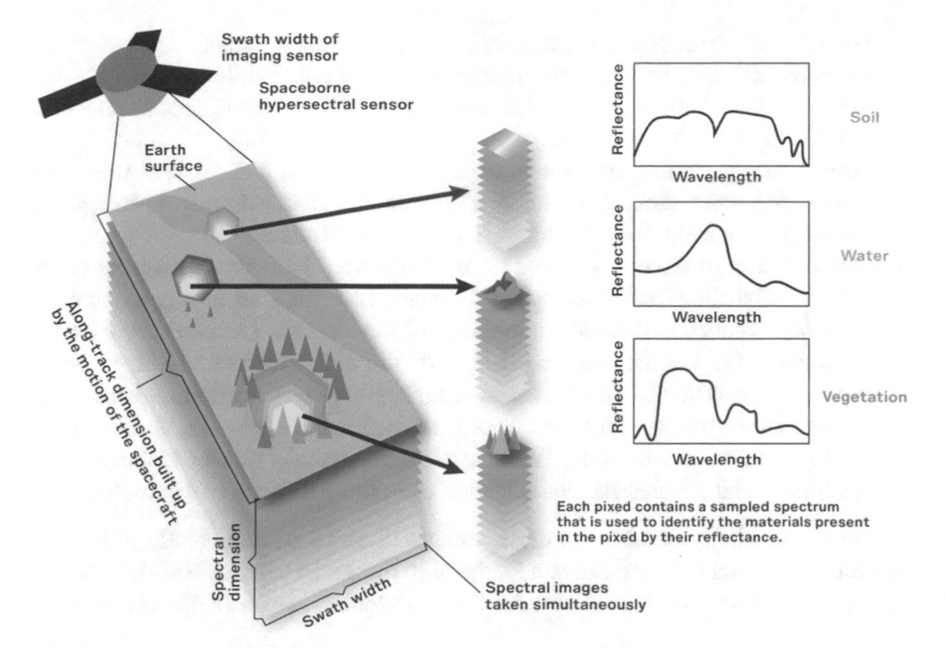

Fig. 1.1 3D cube of HSI data [1]

and sometimes impossible to get the labeled data and if its available, it is very costly and limited. High spectral data and detailed spatial characteristics make HSI analysis a challenging task as it is really strenuous to get the spectral-spatial features without data complexity. Following are some of the challenges:

1. **Effect of noise and other atmospheric factors:** Factors such as illumination and multi-scattering in the acquisition process, noise, and redundancy causes significant difficulties in HSI classification. The analysis of the scenes turns out to be really challenging due to the high spectral redundancy, noise and uncertainty sources observed. Source of illumination may affect the spectral signatures as object illuminated through different sources may have different spectral characteristics at one time.
2. **Curse of dimensionality:** Large ratio between the presence of a huge number of bands in HSI and availability of a small amount of training data makes the analysis of remote sensing scene a challenging task. Methods based on machine learning techniques, this problem makes things even worse as with increase in dimensionality, the amount of labeled data required to attain a statistically reliable result also rise as the volume of parameters to be predicted increases with dimensionality. Optimizing such a large number of parameters become challenging, complex, and time- consuming [5].
3. **Integration of spectral-spatial information:** Recent advancements in remote sensor technology delivers rich spatial information along with the spectral data. Spatial and spectral domains are entirely different, each carries distinctive charac-

teristics and properties. Contextual information comprises edges, object shapes, structures, different textures, and contextual information. While spectral information consists of distinct spectral characteristics of the scene and depicts material-related properties. For an effective and accurate classification, it is recently discovered that exploiting spectral and spatial information plays a vital role. Due to their different nature, it is still an open research area as to how to merge the spectral and spatial information for effective classification.

4. **Classification of Diverse Classes in the Presence of Limited Training data:** Process of labeling each Hyperspectral Image (HSI) pixel according to the class it belongs is called HSI classification. One of the most vital limitations of HSI is the availability of limited training data as its difficult, expensive and sometimes impossible to label the HSI data which makes the classification a challenging task as most of the classifiers and feature extraction techniques require plenty of training data. Moreover, some HSI datasets include many classes within a small size image which makes the classification even more difficult and complex.

It is therefore highly desirable to design a classifier that can efficiently utilize the spectral and spatial data and can handle the high dimensionality of HSI data. Moreover, the detection and removal of redundancy and noise is also an open issue.

1.3 Research Objective

The main purpose of the book is to explore and define novel approaches for the analysis and classification of the hyperspectral remote sensing scenes by incorporating spectral and spatial characteristics of the image. In addition, we develop techniques to effectively extract spatial features in unsupervised manner and design a framework to detect the redundancy and noise in the high-dimensional data. In particular, the prime target is on investigation and optimization of deep learning based deep feature extraction strategies, for the extraction and integration of spectral and spatial characteristics for powerful HSI classification. The framework of the proposed hyperspectral remote sensing image analysis is presented in Fig. 1.2. In recent years, deep learning based architectures, which can extract deep and discriminative features in a hierarchical manner, have gained more attention. Deep learning has recently proved its effectiveness in extracting useful features in HSI classification. However, various problems associated with the computational cost and effective extraction of class-specific statistics need further investigation. The high dimensionality of the HSI data, which contains redundant and noisy information, and which also leads to Hughes phenomena is also an open issue.

In order to address and overcome the above-mentioned challenges and limitations, which obstructs the HSI analysis, the following objectives are established:

- Deeply explore the behavior and performance, in terms of complexity and classification accuracy, of deep learning based architectures/algorithms in the remote sensing field, under various experimental arrangements and based on that design

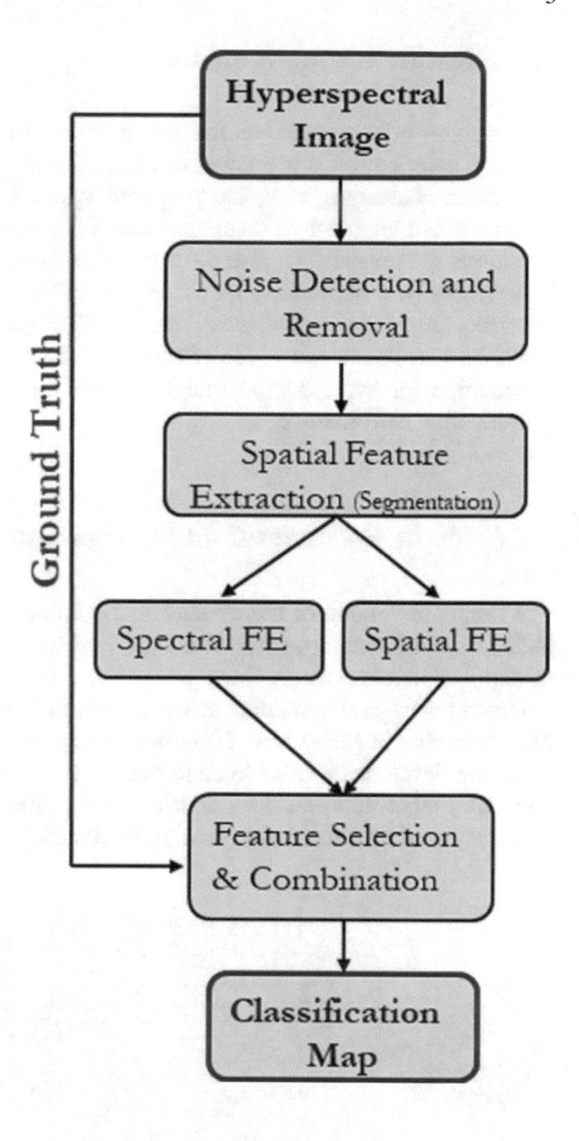

Fig. 1.2 General framework for hyperspectral image analysis

effective deep learning based classification methods for improved classification performance.

- Develop an innovative methodology for spatial feature extraction, information redundancy, and noise issues by exploiting the adaptive boundary adjustment based technique.
- Define an approach for integrating both spectral and spatial characteristics within a deep learning based classification framework.
- Design a novel strategy to address the Hughes phenomenon for HSI classification by exploiting tri-factor-based criteria.

1.4 Research Achievements

This section briefly describes the research achievements presented in this book. Detailed contribution is presented in Figs. 1.3 and 1.4. In order to meet the above-mentioned challenges, first, the proposed spectral feature extraction based band categorization approach is discussed which can effectively detect and remove the redundant and noisy data and at the same time retain the most discriminant information. Secondly, unsupervised spatial feature extraction approach is proposed which segments the spatially similar regions in HSI. Lastly, several deep-learning based algorithms are developed which effectively extract and integrate the spectral-spatial information for hyperspectral image classification which demonstrates improved classification performance.

1.4.1 Noise Reduction/Band Categorization of HSI

The immense volume of the dimensional domain of Hyperspectral Scenes (HSIs) including all of its complexity and intricacy which also involves a substantial amount of duplication and noise is considered to be one of the persistent issue that results in encumbrance and numerous obscurities in HSI analysis and for all the succeeding application in general and HSI classification in particular. To handle the subject issue, we developed a flexible edge detection based group-wise Band Categorization (BC) algorithm that categorizes channels through the content and extent of useful spectral data that exists in a particular channel. Moreover, it also addresses

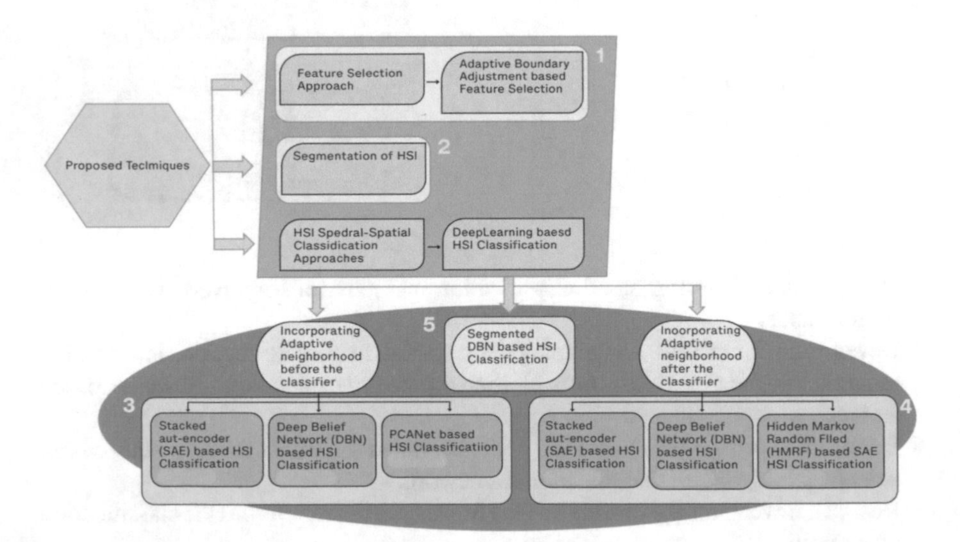

Fig. 1.3 The major contribution of the book

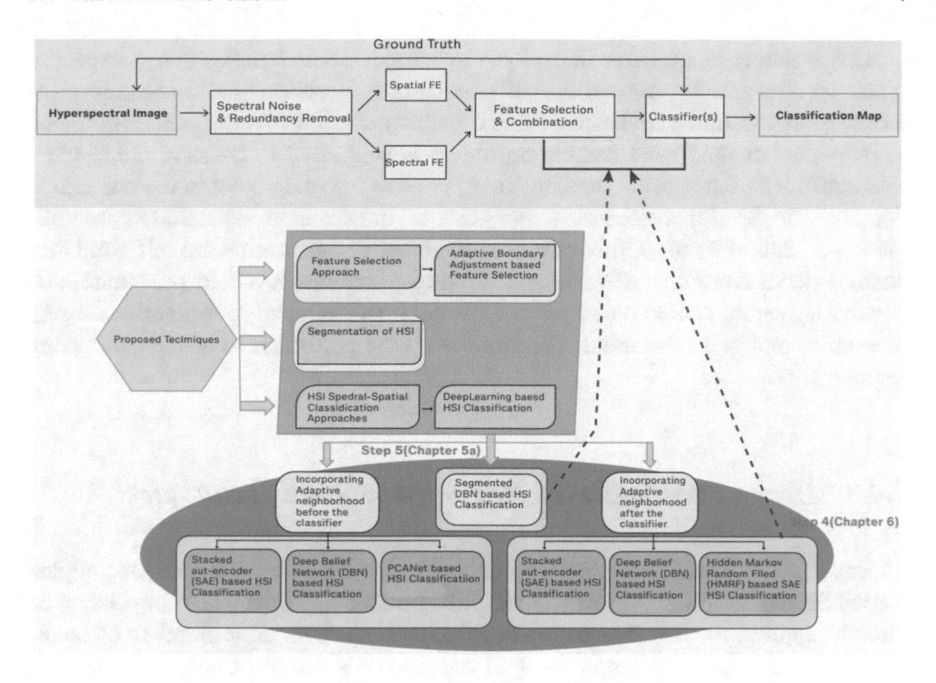

Fig. 1.4 General framework for hyperspectral image analysis along with contribution at each stage

the duplication/noise material deprived of negotiating the real content from the raw hyperspectral data. As hyperspectral scene training data is demanding and complex to acquire, an innovative unsupervised spectral-contextual flexible channel-noise criteria-rooted design is formulated for band categorization which is rooted in capacity of edge adaptation/modification and channel inter-association. Strong clustering and edge-detection-rooted approaches contain two major concerns: channel association and channel preference. Our developed technique anticipates mutually channel correspondence as well as channel synergy. Here, channel preference is established by the distinguishing and information content confined in each spectral channel.

1.4.2 Unsupervised Hyperspectral Image Segmentation

Hyperspectral Image (HSI) segmentation is believed to be the vital preprocessing steps for successive applications and deep analysis to take full advantage of the existing multispectral information. Unsupervised Segmentation devoid of dimensional contraction procedure and without training samples is a persistent issue in computer vision but a more serious and critical issue in hyperspectral imaging. An innovative unsubstantiated spectral-contextual flexible edge modification/alteration rooted architecture and a structure is designed for hyperspectral scene segmentation, which is modified through biased clustering.

This architecture utilizes and explores two major characteristics of hyperspectral scene: spectral relationship and channel contextual preference. Spectral channel preference is demonstrated by the distinctive capability and information content confined in individual channel, and flexible procedure is suggested to preserve the spectral relationships in the spectral domain and to contest the real object in the contextual domain. The developed methodology agrees on the scene to be exploited at multiple segmentation points. Utilization of confined edge uniformity and self-similarity characteristics from channel-cluster flexible edge alteration rooted procedure, and concluding segmentation outcome rooted from the developed four diverse merging measures employ a substantial consequence on the performance of the concluding segmentation.

1.4.3 Deep Learning Based HSI Classification Techniques

DL can characterize and establish various stages of information to represent complex associations between data. However, despite its advantages, deep learning cannot be directly applied on HSI due to multiple reasons such as large number of bands and limited available training data. HSI consists of hundred of bands, hence even a small patch comprises very large data that results in a large number of neurons in a pre-trained network. Similarly, very few labeled samples make the network, difficult to train. Moreover, incorporating spatial contextual information along with the spectral features for DL is also an open research problem. Furthermore, images captured by different sensors exhibit different characteristics and generally present great differences. Therefore, in this book multiple deep learning based classification methods are proposed. In general integrating spatial information along with spectral channels in the HSI classification process is of paramount importance [6] as with the advancement in imaging technology, hyperspectral sensors can deliver an excellent spatial resolution. In this regard, researchers have proposed certain techniques. In the first category, classification techniques extracts the spatial and spectral features before performing classification [7, 8]. In the second category, HSI classification methods consist of incorporating spatial information into the classifier during the classification process [9]. In the third category, classification methods attempt to include spatial dependencies after the classification either by spatial regularization or by decision rule [10]. However, all these spatial features need human knowledge. In this book, several DL-based methods are developed and investigated to effectively incorporate the spatial contextual information in an effective way. These methods can be split into two major groups established on the integration of spatial information and nature of network:

- Integration of spatial information prior to classification

1. Stacked auto-encoder based HSI classification,
2. Deep Belief Network based HSI classification,
3. PCANet based HSI classification.

- Integration of spatial information after classification

  1. Stacked auto-encoder and Hidden Markov Random field based HSI classification,
  2. Deep Belief Network based HSI classification,
  3. segmented DBN based HSI classification.

1.4.3.1 Stacked Auto-encoder Based Spectral-Spatial Classification of HSI

Improving the classification accuracy of diverse classes in HSI is of preeminent concern in remote sensing field. Recently, deep learning algorithms have established their capability in HSI classification. However, despite its learning capability, fixed-size scanning window in deep learning and its inability to integrate spatial contextual information along with spectral features in deep network for improved performance limits its capability. In this work, for spectral-spatial feature extraction, a spatial-adaptive hyper-segmentation-based Stacked Auto-Encoder (SAHS-SAE) approach is proposed, which adaptively modifies the scanning window size and explores spatial contextual features within spectrally similar contiguous pixels for robust HSI classification. The proposed approach includes two key methods–first, we developed adaptive boundary movement based hyper-segmentation whose size and shape can be adapted according to the spatial structures and which consists of spatially contiguous pixels with similar spectral features, second, object-level classification using Stacked Auto-Encoder (SAE) based decision fusion method is developed that integrates spatial-segmented outcome and spectral information into an SAE framework for robust spectral-spatial HSI classification. The proposed approach replaces the traditional scanning window approach for SAE with object-level hyper-segments. Moreover, for robust classification, band preference and correlation-based band selection approach is used to select only the most informative bands without compromising the original content in HSI. Use of local structural regularity and spectral similarity information from adaptive boundary adjustment based process, and fusion of spatial context and spectral features into SAE has a significant effect on the accuracy of the final HSI classification. Experimental results on real diverse hyperspectral imagery with different contexts and resolutions validate the classification accuracy of the proposed method over several well-known existing techniques.

1.4.3.2 Deep Belief Network Based Spectral-Spatial Classification of HSI

Lately, the employment of deep learning based architectures in hyperspectral image analyses has been matured and materialized. However, fusing contextual characteristics along with spectral information in deep learning model is a persistent challenge. This framework represents a distinguishing contextually modified deep belief net

(SDBN) that effectually employs contextual characteristics inside spectrally matching adjacent pixels for hyperspectral scene classification. In the developed framework, scene is initially partitioned into flexible edge regulation rooted contextually comparable areas that possess the same spectral characteristics, succeeding a structural feature mining and classification is commenced utilizing Deep Belief Network (DBN) rooted outcome merging technique that joins contextually partitioned contextual and spectral data into a DBN network for improved spectral-spatial hyperspectral scene classification. Furthermore, for enhanced precision, channel partiality/association rooted characteristic selection technique is utilized to choose the spectral channels with maximum data devoid of conceding the real information in the scene. Employment of indigenous contextual characteristics and spectral correspondence from flexible edge regulation rooted technique, and incorporation of contextual and spectral characteristics into DBN net fallouts into enhanced precision of the concluding scene classification. Experimental demonstration of famous hyperspectral scenes designates the classification precision of the developed approach over numerous prevailing approaches.

1.4.3.3 PCANet Based Spectral-Spatial Classification of HSI

Distribution of each pixel in the HSI scene to a corresponding class by employing feature mining through well-known DL-based architecture has already demonstrated great performance. Nevertheless, the multifaceted net model, wearisome training procedure and active employment of contextual material in deep network bounds the employment and enactment of deep learning. In this portion of the book, for an operative spectral-contextual feature extraction, an improved deep network, contextual flexible network (SANet) technique is developed that employs contextual characteristics and spectral properties to create a further abridged deep network that results in much improved feature mining for the subsequent procedural analysis. SANet is recognized from the effective model of a principal component analysis net. Initially, contextual operational characteristic is mined and fused with useful spectral bands succeeded by a structural classification by utilizing SANet rooted conclusion merging technique. It merges contextual outcome and spectral features into a SANet network for vigorous spectral-contextual scene analyses. A combination of confined operational uniformity and spectral likeness into effective deep SANet has substantial consequences on the classification enactment. Experimental demonstration on prevalent regular HSI scenes exposes the performance of SANet approach that acted much better with increased accuracies.

1.4.3.4 Segmented DBN Based HSI Classification

Deep learning based deep belief networks have lately been designed for feature extraction in hyperspectral scenes. Deep belief net, as deep learning based architecture, has been utilized in hyperspectral scene analyses for shallow and invariant features extraction. Nevertheless, DBN architecture has to face and handle numerous spectral characteristics and high spatial resolution from hyperspectral cube, which leads to the intricacy and inability to mine true exact invariant characteristics, hence the ability of this DL architecture damages badly in front of the hyperspectral challenges. Furthermore, dimensionality reduction based solution to the subject problem results in damage of valued spectral data, which further lowers the accuracy. To handle this issue, this section develops a spectral-variational segmented DBN (SAS-DBN) for spectral-contextual hyperspectral classification that explores the invariant deep characteristics by partitioning the real spectral channels into tiny groups of associated spectral channels and applying deep belief net to each individual group of channels independently. Additionally, contextual characteristics are also merged by initially employing hyper-segmentation on the scene. The performance of this approach improved the classification accuracy as expected. By indigenously employing DBN-rooted characteristics mining to every individual channel group decreases the computational intricacy and simultaneously leads to improved data mining and, therefore, enhanced precision is acquired. Overall, employing spectral characteristics effectually through partitioned DBN procedure and contextual characteristics by flexible-segmentation and addition of spectral and contextual characteristics for scene analyses made a foremost impact on the accuracy of classification. Experimental analyses of the developed approach on prevailing hyperspectral typical scenes with diverse contextual features and resolutions launch the worth of the developed approach where the outcome is similar to numerous newly developed hyperspectral classification approaches.

1.4.3.5 Stacked Auto-encoder and Markov Random Field Based HSI Classification

This technique develops a novel spectral-contextual hyperspectral scene classifying methodology built on invariant characteristics mining by utilizing Stack-Auto-Encoders (SAE) along with unsupervised hyperspectral segmentation. Precisely, initially, the SAE architecture is employed as a standard spectral feature-rooted classifier for invariant characteristic mining. Subsequently, contextual subjugated feature is obtained by utilizing operative edge regularization focused segmentation approach. Lastly, the supreme voting based feature is employed to fuse the spectral mined characteristics and contextual associations, that forms a precise classification map.

1.5 Organization of the Book

The rest of the book is arranged as follows: Chap. 2 briefly provides a description of hyperspectral imaging, and several deep learning based classification techniques for hyperspectral image classification. Chapter 3 presents the proposed boundary adjustment based band selection/categorization approach for effective spectral feature extraction. Moreover, Chap. 4 includes the proposed approach for spatial feature extraction through Adaptive boundary adjustment based criteria. Chapter 5 introduces a novel concept of adaptive window size and spatial feature fusion for optimized deep learning feature extraction. It presents a new methodology that integrates the findings of Chaps. 3 and 4, by integrating the spectral and spatial information in a deep learning architecture for HSI classification. Chapter 6 presents the strategies for incorporating the spatial information and exploiting the spectral channels for HSI classification. Chapter 7 presents the sparse-based deep learning solution of HSI classification. Finally, Chap. 8 summarizes this book.

References

1. http://www.markelowitz.com/Hyperspectral.html
2. Marin-Franch I, Foster DH (2013) Estimating information from image colors: an application to digital cameras and natural scenes. IEEE Trans Pattern Anal Mach Intell 35(1):78–91
3. Kim SJ, Deng F, Brown MS (2011) Visual enhancement of old documents with hyperspectral imaging. Pattern Recognit 44(7):1461–1469
4. Fu Z, Robles-Kelly A, Zhou J (2011) MILIS: multiple instance learning with instance selection. IEEE Trans Pattern Anal Mach Intell 33(5):958–977
5. Hughes G (1968) On the mean accuracy of statistical pattern recognizers. IEEE Trans Inf Theory 14(1):55–63
6. Plaza A, Plaza J, Martin G (2009) Incorporation of spatial constraints into spectral mixture analysis of remotely sensed hyperspectral data. In: IEEE international workshop on machine learning for signal processing, 2009. MLSP 2009. IEEE, pp 1–6
7. Li J, Marpu PR, Plaza A, Bioucas-Dias JM, Benediktsson JA (2013) Generalized composite kernel framework for hyperspectral image classification. IEEE Trans Geosci Remote Sens 51(9):4816–4829
8. Zhou Y, Peng J, Chen CP (2015) Extreme learning machine with composite kernels for hyperspectral image classification. IEEE J Sel Top Appl Earth Obs Remote Sens 8(6):2351–2360
9. Chen Y, Nasrabadi NM, Tran TD (2013) Hyperspectral image classification via kernel sparse representation. IEEE Trans Geosci Remote Sens 51(1):217–231
10. Fauvel M, Benediktsson JA, Chanussot J, Sveinsson JR (2008) Spectral and spatial classification of hyperspectral data using SVMS and morphological profiles. IEEE Trans Geosci Remote Sens 46(11):3804–3814

Chapter 2
Hyperspectral Image and Classification Approaches

2.1 Introduction to Hyperspectral Imaging

Recent developments in remote sensing technology and geographical data have directed the way for the advancement of hyperspectral sensors. Hyperspectral Remote Sensing (HRS), also known as imaging spectroscopy, is a comparatively new technology that is presently under investigation by researchers and scientists for its vast range of applications such as target detection, minerals identification, vegetation, and identification of human structures and backgrounds. HRS integrates imaging and spectroscopy in a distinct structure that generally consists of huge data sets and needs a modern state-of-the-art analysis techniques. Electromagnetic spectrum of light is shown in Fig. 2.1. Hyperspectral images mostly consist of spectral channels in the range of about 100–200 in the narrow bandwidth range of 5–10 nm, while, multispectral images generally consist of 5–10 spectral channels in large bandwidth range, i.e., 70–400 nm.

2.1.1 Hyperspectral Imaging System

When the light interacts with the earth's surface, 5 mechanisms can happen, either the light can scatter in many directions, reflect in a single direction, absorbed as a energy and stored in that material or transmitted or passes through. Figure 2.2 presents a detailed process. If we only consider the reflection component, the reflection of sun's energy by any earth material creates a distinct footprint specifically known as the spectral signature of that particular material. The location and shape of these unique spectral signatures enable us to identify the different types of the land surface features.

L. Tao and A. Mughees, *Deep Learning for Hyperspectral Image Analysis and Classification*, Engineering Applications of Computational Methods 5,
https://doi.org/10.1007/978-981-33-4420-4_2

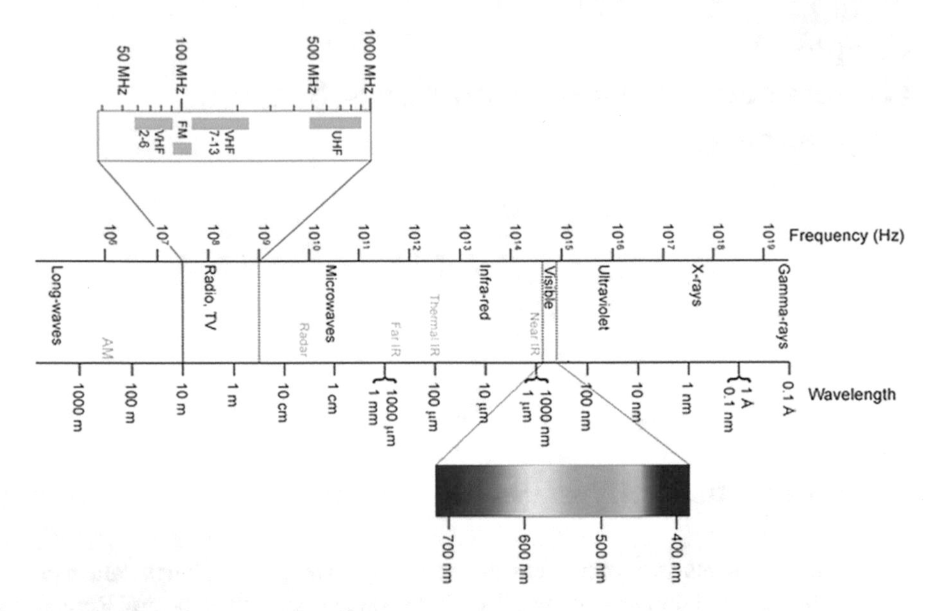

Fig. 2.1 Electromagnetic spectrum [1]

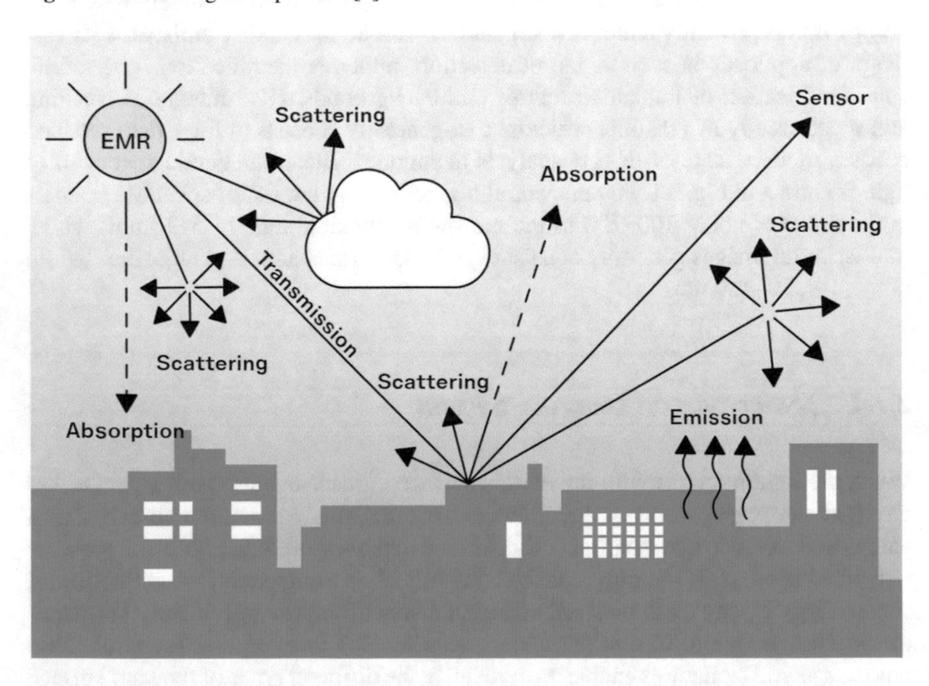

Fig. 2.2 Electromagnetic radiation's interaction in the atmosphere and Earth's surface [2]

2.1.1.1 Bands and Wavelengths

Each of the bands in the 200 bands of HSI is associated with the particular wavelength region. For instance, the wavelength region between 700 and 705 nm might be associated with one band as captured by an imaging spectrometer. Spectral resolution is associated with a number of bands. But it is not only the number of bands but also band range (bandwidth). Smaller the bandwidth, better is the spectral resolution and vice versa. This is the main difference in multispectral and hyperspectral, i.e., spectral resolution is higher in hyperspectral than in narrow-band observation.

Hyperspectral imaging is a process of collecting and processing of information across the electromagnetic spectrum with hundreds of spectral bands. Figure 2.3 presents the difference between broadband, multispectral, hyperspectral, and ultraspectral (spectral sensing). First is panchromatic, the entire visible region falls into one observation. In multispectral, the same visible region is broken down into more broad spectral channels. Whereas, in case of hyperspectral image, the same region is divided into hundreds of bands. As the spectral capability increases, the power of analysis also increases, the kind of detailed information that you can retrieve also increases. In hyperspectral image, for each image, you have a large number of observations depending on the number of bands you have through which you can identify the ground material by matching its spectral signature.

2.1.2 Why Hyperspectral Remote Sensing

Most of the earth surface materials have diagnostic absorption features in the 400–2500 nm range of the electromagnetic spectrum. These features are of a very narrow spectral appearance. These materials can only be identified if the spectrum is sampled at sufficiently high spectral resolution.

Human eye can only detect the reflective energy in the visible region of the electromagnetic spectrum, i.e., 0.4–0.7 μm. On the other hand, advanced hyperspectral sensors capture data in the form of image, sensing a part of the electromagnetic radiation reflected from the Earth's surface in a range of wavelengths including visible, near-infrared, and short-wavelength infrared regions of the electromagnetic spectrum as described in Fig. 2.4.

2.2 Review of Machine Learning Based Approaches for Hyperspectral Image Classification

In order to classify, identify, and analyze the chemical composition of objects in the area, hyperspectral acquiring devices enable scene contents to be remotely analyzed. Hyperspectral images of earth observation satellites and aircraft have therefore been

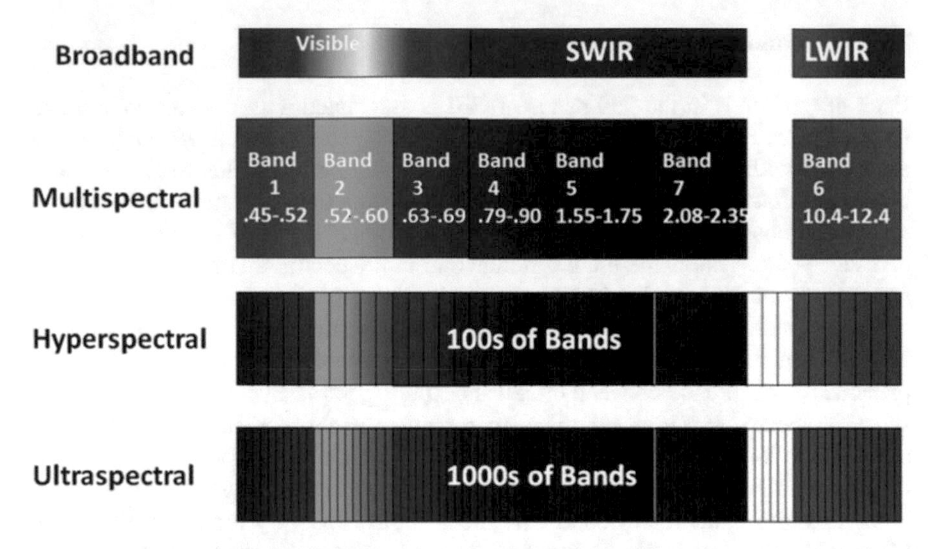

Fig. 2.3 Difference between multispectral and hyperspectral data [3]

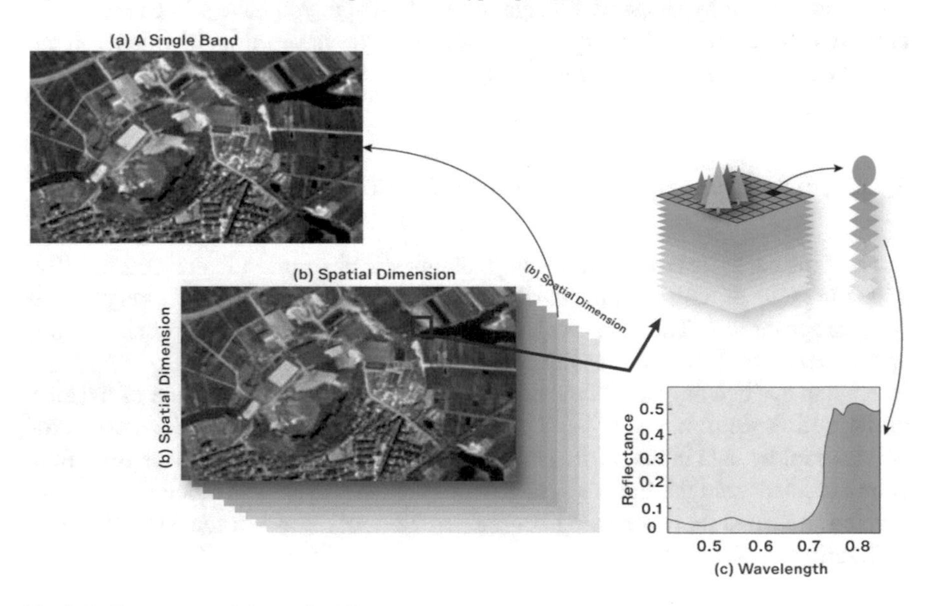

Fig. 2.4 Hyperspectral data cube [4]

increasingly important for agriculture, environmental surveillance, urban development, mining and defense purposes. Hyperspectral scenes of earth observation satellites and aircraft have therefore been increasingly important for agriculture, environmental surveillance, urban development, mining, and defense purposes. Machine Learning (ML) algorithms have become a crucial method for modern hyperspectral imaging research due to their extraordinary predictive ability. For remote sensing scholars and scientists, thus a sound knowledge of ML techniques has become extremely important. This section of the chapter examines and analyzes newly developed algorithms in the research community for ML learning based hyperspectral scene classification. These techniques are organized by scene evaluation as well as the ML algorithm and produce a dual visualization of the scene assessment and different kinds of machine learning based scene analysis algorithms. Both hyperspectral images interpretation and ML techniques are covered in this section (Fig. 2.4).

In the areas such as agricultural production, environmental sciences, wildlife, mineral extraction and rapid urbanization, security and aerospace science, hyperspectral imagery is considered as a good remotely sensed tool for studying the chemical structure of earth resources. Remote sensing, sometimes regarded as spectroscopy image acquisition, measures the transmitted or released electromagnetic radiation throughout the image from visible to infrared frequencies into multiple of hundreds of associated spectral relations. Every pixel in a hyperspectral scene has a vector consisting of hundreds of components, the scale factor is recognized as the spectral range, that measure the reflective or emitting radiation. Hyperspectral images can therefore be perceived as a 3-dimensional data model comprising of 2 contextual axes that carry relevant data regarding object placement and a spectral axis that carries data regarding the chemical texture of elements. This frequency range acquires chemical features as these atoms or molecule structures are controlled by the interface among light at distinct frequencies and components. Aerial and satellite-based hyperspectral devices can capture scenes and also can design the earth surface and earth usage, identify and localize structures, or interpret the substances' physiological characteristics over a broad geographic region. Hyperspectral scenes, which consist of hundreds of channels, cannot be evaluated just like color scenes as there are only 3 channels in RGB scenes. Hence, computer vision approaches are designed to obtain expressive data from the scenes. ML and computer vision rooted techniques have demonstrated their accuracy in this regard, because they have the capability to inevitably pick the association among the spectral information acquired at each spatial location in the scene and the characteristics which are required to be acquired. In terms of managing distortion and ambiguities, they possess much more robustness in comparison to conventional approaches, like hand-engineered standardized indices and physics rooted architectures. The practitioners associated with the hyperspectral field has revealed a prodigious curiosity in ML in general and in deep learning in specific.

This section targets to deliver a wide-ranging exposure of not only hyperspectral scene classification job but also ML and specifically deep learning algorithmic views. All the approaches summarized in this section are mature published studies. These approaches are capable of analyzing electromagnetic emission as well as reflective

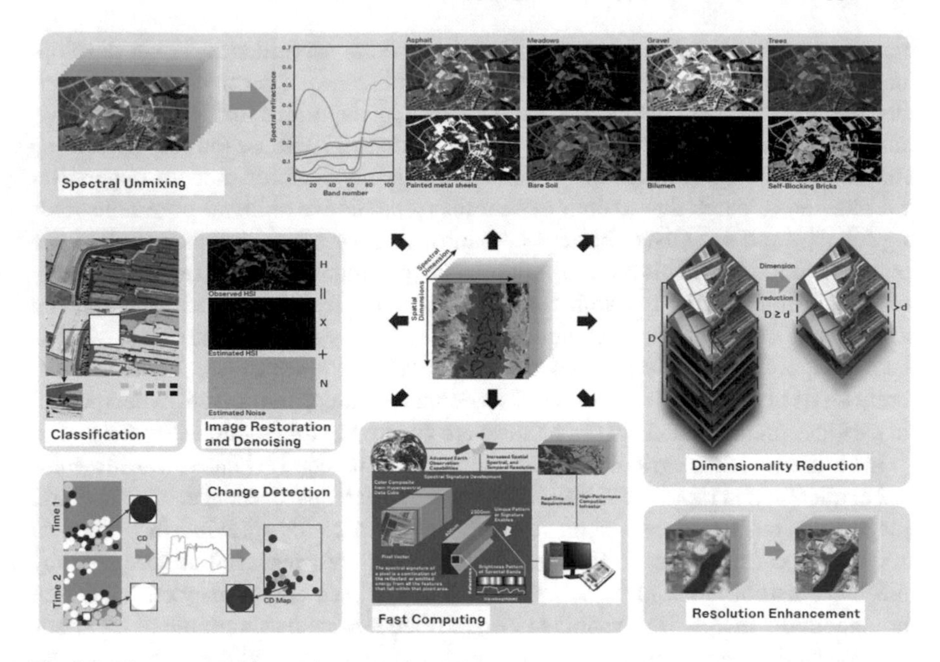

Fig. 2.5 Hyperspectral image interpretation [5]

scenes, except presented the other way around. The hyperspectral image exploration goal is characterized as earth surface classification [39], object localization [222], unmixing [29], and somatic factor assessment [278]. The major aim of this section is to get the reader familiarized with newly researched HSI classification techniques, classify a technique either by hyperspectral image job or an ML algorithm and investigation of existing tendencies and challenges including upcoming directions.

2.2.1 Hyperspectral Image Interpretation Taxonomy

The attribute of the surface component which governs the degree of the reflective energy is the angular reflection of the surface. Nevertheless, the energy entering the receptor involves inputs from the atmospheric dispersion, which can be extracted by utilizing the environmental adjustment approaches [101] for the estimation of the surface reflection. Therefore, the pixel values in the hyperspectral scene are evaluated at the radiance or reflective level. The reflective characteristic of a scene is further favored for HSI scene exploration due to the surface characteristic of the reflective attribute as shown in Fig. 2.5. Moreover, the scene containing reflective attributes generates enhanced results due to a reduction in the environmental intervention. Hyperspectral scenes can be analyzed in 4 discrete fields: land cover mapping, target

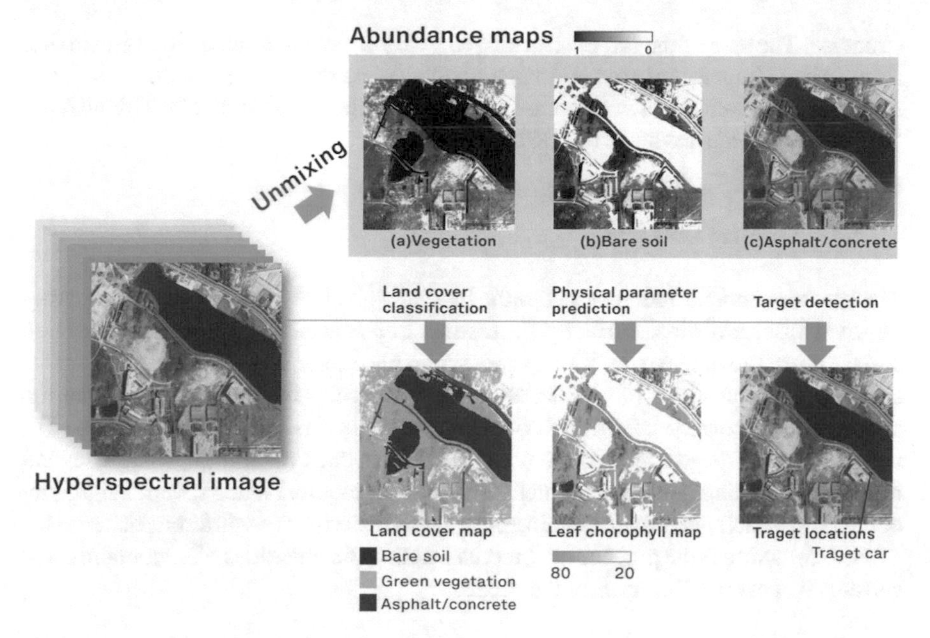

Fig. 2.6 Hyperspectral image interpretation [5]

identification, spectral unmixing, and physiological factor assessment, as depicted in Fig. 2.6.

2.2.1.1 Land Surface Mapping

Earth surface mapping [39] is the procedure of detecting the substance to which each pixel of the HSI scene belongs to. The objective is to generate a map presenting the diverse distribution of several substances over a terrestrial region captured by an HSI device. Important uses of surface mapping includes plant types taxonomy [69], city image organization [74], mineral detection [218], and variation investigation [238].

Numerous surface area mapping techniques involve a preceding information about the categories of substances that exist in the image including the spectral signature that belong to that particular substance. In general, this data is supplied by professionals from pixel values obtained from the field, or altered from a given spectrum collection. Nevertheless, several surface area mapping approaches do not need preceding data for the image substance.

2.2.1.2 Target Identification

The role of target identification [191] in a hyperspectral scene is to identify and localize the destination structures provided the spectral signature of a particular structure. The size of the target structure can vary from a few pixels to even smaller than a pixel. Targets that are less than a pixel size are difficult to detect.

Task linked to target identification is anomaly detection that involves identifying the unusual objects resent in the HSI scene.

2.2.1.3 Spectral Unmixing

The electromagnetic radiance acquired by each pixel of the HSI scene is hardly returned from a distinct surface of a distinct substance. Scenes captured through aerial or satellite have a resolution of more than one meter, i.e., each pixel represents an area which mostly is more than one meter. Hence, it is highly possible that the particular area consists of different or sometimes numerous different materials. For instance, in an image captured from an urban area, each pixel may contain several materials including man-made structures, roads, trees, etc. Hence spectral signature obtained for each pixel may contain spectral characteristics of different materials.

HSI unmixing is the procedure of reconstructing the quantities of uncontaminated substance at every pixel level of the scene.

2.3 Hyperspectral Remote Sensing Image Dataset Description

To evaluate the classification performance, all the researchers [6] in the HSI classification research community utilizes these available standard real hyperspectral data sets captured by different sensors at different times at different locations. These datasets propose challenging classification tasks due to the presence of both rural and urban areas as well as small man-made and natural structures. Mostly, in the existing literature, two datasets at each time, are considered to demonstrate the validation and accuracy of the proposed techniques for HSI classification. A detailed description of each dataset in tabular form is given in Table 2.1. A brief description of each dataset is given below.

2.3.1 Indian Pine: AVIRIS Dataset

Indian Pine dataset was acquired through Airborne Visible Infrared Imaging Spectrometer (AVIRIS) sensor in 1992 over the pines region of Northwestern Indiana. It consists of spatial size of 145×145 with a ground resolution of 17 m. It contains 224 spectral bands in the wavelength range 0.4–2.5 m, with a spectral resolution of 10 nm and a spatial resolution of 20 m. Out of 224, 24 noisy bands due to water absorption were removed resulting in 200 spectral channels. As shown in Fig. 2.7,

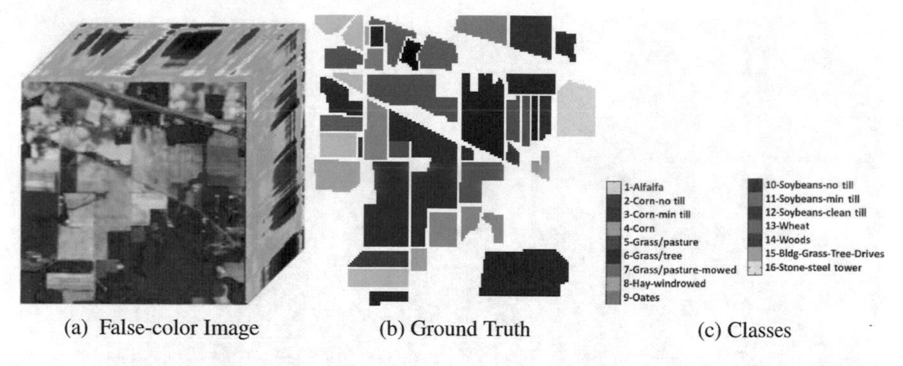

(a) False-color Image (b) Ground Truth (c) Classes

Fig. 2.7 Indian Pine dataset with 16 classes

it contains 16 different land cover agricultural classes. False-color composition and ground truth are presented in Fig. 2.7. This dataset is considered to be one of the challenging datasets due to its low spatial resolution, small structural size, and presence of mixed pixels.

2.3.2 Pavia University: ROSIS Dataset

The Pavia University Scene was collected by Reflective Optics System Imaging Spectrometer (ROSIS) sensor over Pavia University, Italy. The Pavia scene comprises of a spatial size of 610×340 and a spectral size of 115 channels. The spatial size of the scene is 1.3 m/pixel while the spectral range is 0.43–0.86 μm. A total of 12 noisy bands were removed owing to water absorption with 103 remaining bands. Nine standard classes are utilized for Pavia scene classification. The false-color composite is described in Fig. 2.8. This dataset comprises both man-made structures and green areas.

2.3.3 Houston Image: AVIRIS Dataset

The Houston database was collected by AVIRIS sensor over the University of Houston, and neighboring urban region. It contains 144 spectral bands with a spectral resolution of 380×1050 nm and a spatial area of 349×1905. It consists of 15 different ground cover classes as shown in the false-color composite and ground truth in Fig. 2.9.

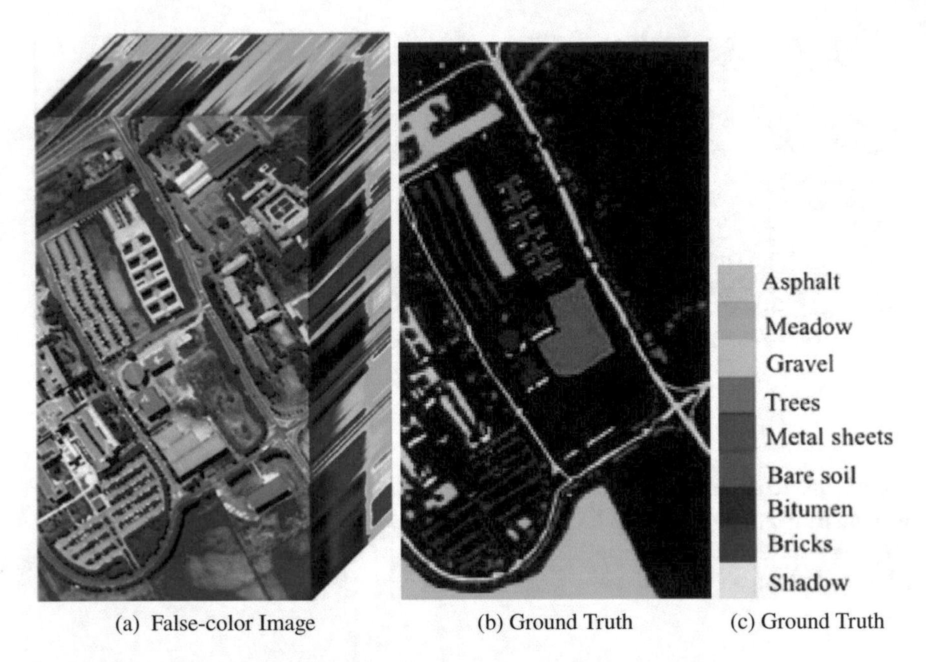

(a) False-color Image (b) Ground Truth (c) Ground Truth

Fig. 2.8 Pavia University dataset with 9 classes

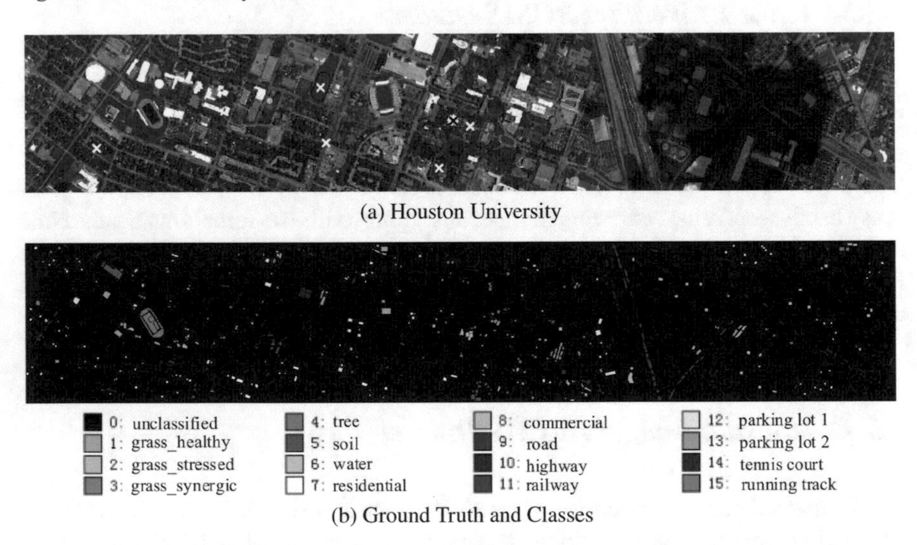

(a) Houston University

(b) Ground Truth and Classes

Fig. 2.9 Houston University dataset with 15 classes

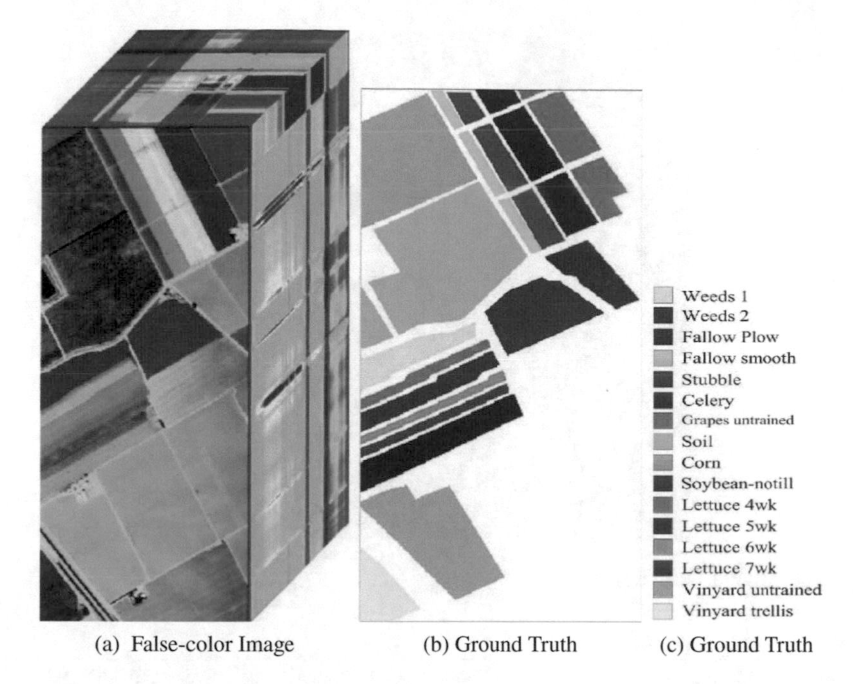

(a) False-color Image (b) Ground Truth (c) Ground Truth

Fig. 2.10 Salinas dataset with 16 classes

2.3.4 Salinas Valley: AVIRIS Dataset

Salinas Valley image was captured by the AVIRIS sensor over the Salinas Valley, California. It comprises of 224 channels originally but 20 water absorption bands were discarded. It is characterized by a high spatial resolution of 3.7 m/pixel. Salinas Valley ground truth consists of 16 classes including soil, green fields. The area covered comprises 512 lines by 217 samples. The terrain and boundaries are shown in Fig. 2.10.

2.3.5 Moffett Image: AVIRIS Dataset

Moffett was acquired in 1997 by the AVIRIS sensor over Moffett Field, at the southern end of the San Francisco Bay, California.[1] The image depicted in Fig. 2.11 has 224 spectral bands, a nominal bandwidth of 10 nm, and a dimensional of 512×614. It consists of a part of a lake and a coastal area composed of vegetation and soil.

[1] [Online] http://aviris.jpl.nasa.gov/html/aviris.freedata.html.

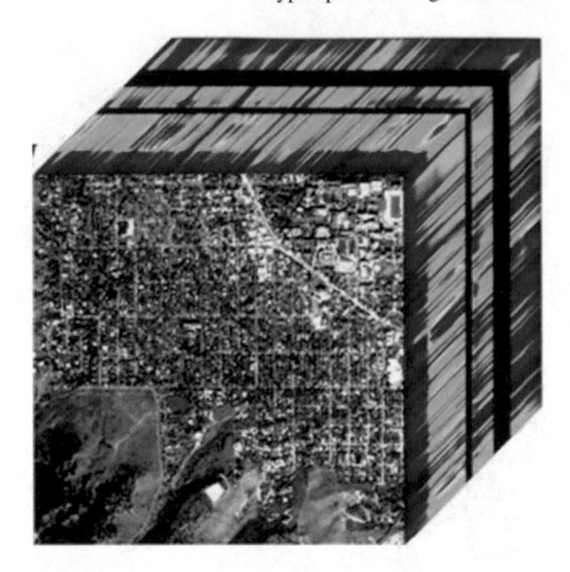

Fig. 2.11 Moffett Field dataset

Table 2.1 Dataset specifications

Dataset	Image size	Classes	Bands	Labeled pixels	Wavelength range (nm)	Spatial resolution (m)
Indian Pine	145 × 145	16	200	10,249	400–2,500	20
Pavia University	340 × 610	9	103	42,776	430–860	1.3
Salinas	512 × 217	16	200	54,129	400–2,500	3.7
Houston University	349 × 1905	15	144	2.5	380–1,050	–
Moffett Field	512 × 614	–	224	–	–	–

2.3.6 *Washington DC Mall Hyperspectral Dataset*

This dataset was acquired by utilizing a Hyperspectral Digital Imagery Collection Experiment (HYDICE) sensor. It comprises 210 spectral channels in the series of 0.4–2.4 um. However, 19 channels were rejected because of water absorption. This hyperspectral image comprises of 1280 rows and 307 columns. In the experiment, 7 material classes were utilized that include Rooftop, Road, Trail, Grass, Tree, Water, and Shadow as shown in Fig. 2.12. The subject HSI is comparatively not as problematic as AVIRIS in terms of classification.

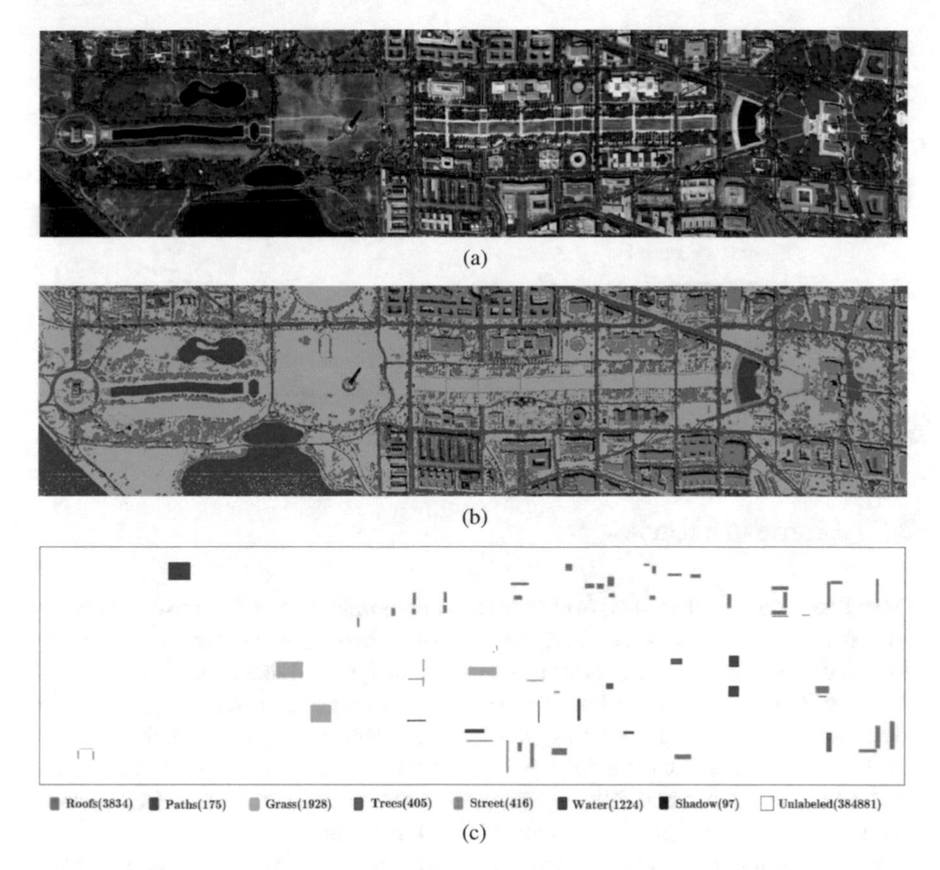

Fig. 2.12 Washington DC Mall dataset. **a** False-color representation of Washington DC. **b** Ground truth. **c** Classes

2.4 Classification Evaluation Measures

In the hyperspectral Image analysis research community, classification performance is estimated using the evaluation criterion's based on overall accuracy (OA), Average accuracy (AA), and kappa Coefficient (k) (Fig. 2.12).

- **Overall accuracy (OA):** OA is the percentage of pixels correctly classified.
- **Average accuracy (AA):** AA is the mean of all the class-specific accuracies over the total number for classes for the specific image.
- **Kappa Coefficient (κ):** Kappa (κ) is a degree of agreement between predicted class accuracy and reality. Generally, it is considered more robust than OA and AA.

A detailed definition of each evaluation measure is presented in Fig. 2.13.

Symbol	Name	Definition
d	Number of classes	
n_i	Number of labelled samples in class i (ground label)	
m_i	Number of classified samples in class i (classification label)	
N	Total number of samples	$N = \sum_{i=1}^{d} n_i = \sum_{i=1}^{d} m_i$
c_i	Number of true positive samples in class i	
p_o	Observed agreement	$p_o = \frac{1}{N}\sum_{i=1}^{d} c_i$
p_c	Chance agreement	$p_c = \frac{1}{N^2}\sum_{i=1}^{d} m_i n_i$
OA	Overall accuracy	$\mathrm{OA} = p_o$
AA	Average accuracy	$\mathrm{AA} = \frac{1}{d}\sum_{i=1}^{d} c_i/m_i$
Kappa	κ	$\kappa = (p_o - p_c)/(1 - p_c)$

Fig. 2.13 Definition and evaluation measures for classification performance

2.5 Literature Review

Deep Learning (DL) based algorithms extract discriminative and representative features from the complex data recently outperformed many typical algorithms in many areas including audio, video, speech, and image analysis. DL has recently been introduced in Remote Sensing and has demonstrated convincing results. By feeding the DL network with the spectral data, the features from the top layer of the network are fed into a classifier for pixel-wise classification. Moreover, by addressing the HSI-related issues and carefully adjusting the input–output parameters, we observed that DL can perform significantly better in HSI analysis.

In this section, the overall framework of DL for HSI classification is presented, along with the advanced DL-based deep networks and tuning tricks. How useful features can be extracted from such an increased volume of data. Traditional feature extraction approaches extracts the feature through models [7]. Substantial development has been attained in recent years, in HSI classification such as handcrafted feature [8–10], discriminative feature learning [11–13], and classifier designing [14, 15]. However, most of the existing techniques can extract only shallow features, which is not strong enough for the classification task. Moreover, these approaches are unable to extract the deep discriminative and representative feature due to the requirement of handcrafted features [11, 16]. Even these handcrafted features don't contain the details of the complex data. The problem even worsens due to the great variability of the HSI data. Detailed description of the existing approaches is depicted in Fig. 2.14 [4]. Thanks to deep learning architectures [17], which provides deep, shallow, discriminative, and representative features for HSI classification. Even though deep learning contains complex and diverse hierarchical architectures, DL methods for HSI classification can be integrated into one broad framework. A general framework of DL methods for HSI analysis is presented in Fig. 2.15.

It comprises three main phases, preprocessed input data, hierarchical multi-layer core deep model, and the extracted output features and classification. In the first

Criteria	Types	Brief Description
Whether training samples are used or not?	Supervised classifiers	Supervised approaches classify input data using a set of representative samples for each class, known as training samples.
	Unsupervised classifiers	Unsupervised approaches, also known as clustering, do not consider the labels of training samples to classify the input data.
	Semi-supervised classifiers	The training step in semi-supervised approaches is based on both labeled training samples and unlabeled samples.
Whether any assumption on the distribution of the input data is considered or not?	Parametric classifiers	Parametric classifiers are based on the assumption that the probability density function for each class is known.
	Non-parametric classifiers	Non-parametric classifiers are not constrained by any assumptions on the distribution of input data.
Either a single classifier or an ensemble classifier is taken into account?	Single classifier classifiers	In this type of approaches, a single classifier is taken into account to allocate a class label for a given pixel.
	Ensemble (multi) classifier	In this type of approaches, a set of classifiers (multiple classifiers) is taken into account to allocate a class label for a given pixel.
Whether the technique uses hard partitioning, in which each data point belongs to exactly one cluster or not?	Hard classifiers	Hard classification techniques do not consider the continuous changes of different land cover classes from one to another.
	Soft (fuzzy) classifiers	Fuzzy classifiers model the gradual boundary changes by providing measurements of the degree of similarity of all classes.
If spatial information is taken into account?	Spectral classifiers	This type of approaches consider the hyperspectral image as a list of spectral measurements with no spatial organization.
	Spatial classifiers	This type of approaches classify the input data using spatially adjacent pixels, based on either a crisp or adaptive neighborhood system.
	Spectral-spatial classifiers	Sequence of spectral and spatial information is taken into account for the classification of hyperspectral data.
Whether the classifier learns a model of the joint probability of the input and the labeled pixels?	Generative classifiers	This type of approaches learns a model of the joint probability of the input and the labeled pixels, and makes the prediction using Bayes rules.
	Discriminative classifiers	This type of approaches learns conditional probability distribution, or learns a direct map from inputs to class labels.
Whether the classifier predicts a probability distribution over a set of classes, given a sample input?	Probabilistic classifiers	This type of approaches is able to predict, given a sample input, a probability distribution over a set of classes.
	Non- probabilistic classifiers	This type of approaches simply assign the sample to the most likely class that the sample should belong to.
Which type of pixel information is used?	Sub-pixel classifiers	In this type of approaches, the spectral value of each pixel is assumed to be a linear or non-linear combination of endmembers (pure materials).
	Per-pixel	Input pixel vectors are fed to classifiers as inputs.
	Object- based and Object- oriented classifiers	In this type of approaches, a segmentation technique allocates a label for each pixel in the image in such a way that pixels with the same label share certain visual characteristics. In this case, objects are known as underlying units after applying segmentation. Classification is conducted based on the objects instead of a single pixel.
	Per-field classifiers	This type of classifiers is obtained using a combination of RS and GIS techniques. In this context, raster and vector data are integrated in a classification. The vector data are often used to subdivide an image into parcels, and classification is based on the parcels.

Fig. 2.14 Summary of classification approaches [4]

phase, the input vector comprises of either spectral feature vector, spatial feature vector, or spectral-spatial feature vector combined.

2.5.1 HSI Noise/Redundancy Detection

Increased volume of HSI data cubes frequently covers a large amount of redundancy and noise that has some undesirable statistical and geometrical characteristics. This

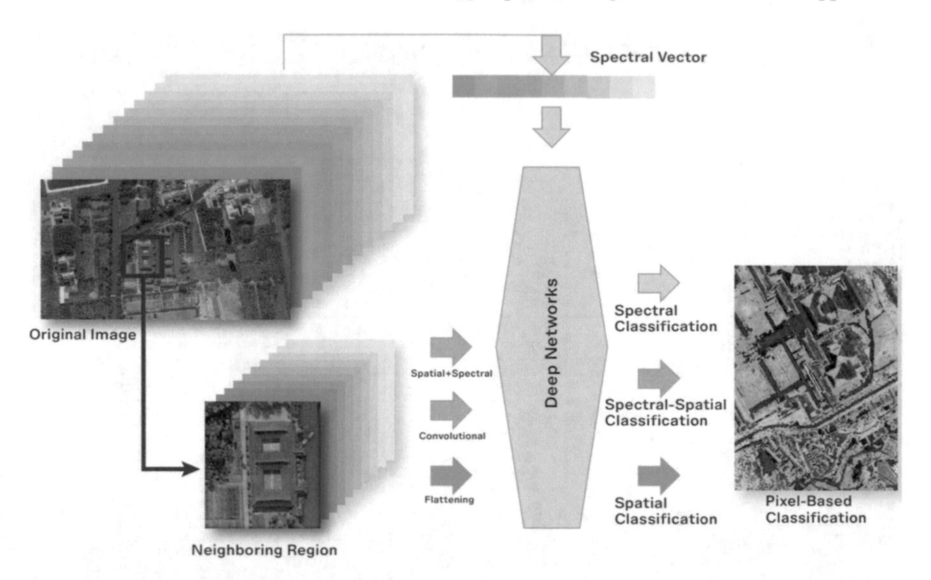

Fig. 2.15 A general framework for the pixel classification of hyperspectral images using DL methods [18]

drawback is due to a number of reasons such as sensor or instrumental noise, environmental effects. Sensor noise comprises thermal noise, quantization noise, and shot Noise, which is the basis of degrading and corrupting the spectral data.

The capability to evaluate the noise features of hyperspectral remote sensing image is a vital stage in creating its abilities and limitations in an operative and scientific perspective. Many efforts have been made to detect and remove noise and redundant data. Noise in HSI can be grouped into two main classes [19]: random noise and fixed-pattern noise. Random noise, due to its stochastic nature, cannot be removed easily. HSI-processing algorithms are usually based on specific noise models, and their performance may reasonably degrade if the model does not properly address the noise characteristics. A widely used random noise model in HSI is the additive model [20, 21]. Most of the hyperspectral processing literature does not address any atmospheric effects that were not mitigated through other methods within their noise models. HIS-processing encounters noise contamination before the radiance enters the sensor similar to water-vapor absorption.

Existing noise-estimation methods can mainly be classified into three types: block-based, filter-based, and their combination. Block-based approaches first divide an image into dense blocks, followed by the discovery of blocks with the least structure and texture, then estimation of noise variance based on these blocks. Commonly, these approaches use variance, spatial homogeneity or spatiotemporal homogeneity [22, 23], local-uniformity analyzers [24], and gradient-covariance matrices [25] as a measurement of structure.

Band Selection (BS) work can be grouped into two categories [26]: (1) Maximum Information or Minimum Correlation (MIMC)-based techniques and (2) Maximum Interband Separability (MIS)-based techniques. MIMC techniques typically use intraband-correlation and cluster criterion. The intraband-correlation criterion-based algorithm gathers suitable subsets of bands by maximizing the overall amount of information using entropy-like measurements [27, 28]. MIS-based algorithms select the suitable set of bands having minimum intraband correlations. For example, [29, 30] included a mutual information-based algorithm and a Constrained Band Selection algorithm based on Constrained-Energy Minimization (CBS-CEM), respectively.

2.5.2 Deep Learning Based Algorithms

In recent years, several DL-based algorithms have been presented [31] and have outperformed existing techniques in many fields such as audio identification [32], natural language processing [33], image classification [34, 35]. The motivation for such an idea is inspired by multiple levels of abstraction human brain for the processing of tasks such as objection identification [36]. Motivated by the multiple levels of abstraction and depth of the human brain, researchers have established innovative deep architectures as an alternative to traditional shallow architectures.

2.5.2.1 Auto-encoder

It is a feed-forward neural network, much like a typical neural network for classification but the main difference is its objective to replicate the input onto the output layer unlike feed-forward, where the objective is to characterize a sharing of a particular class at the output layer. Figure 2.16 shows an auto-encoder with a single hidden layer. Input and output layers in auto-encoder have the same size. During training, we compare the values at the output produced by the auto-encoder with the input data and encourage the auto-encoder to reproduce as perfectly as possible, at the output layer, the values which are at the input layer. Other than this, it is a regular neural network. In auto-encoder, the part of the model that computes the hidden layer is called encoder, which encodes the input into latent representation:

$$l = f(w_l x + b_l), m = f(w_m y + b_m) \tag{2.1}$$

The major aim is to reduce the disparity among the input and the output:

$$\underset{w_l, w_m . b_l, b_m}{\arg\min} \; [error(l, m)] \tag{2.2}$$

For encoding, a typical sigmoid of the linear transformation is utilized. On the other hand, decoder, which is going to take the latent representation h(x), i.e., output of the encoder, linear transform it, and pass it through nonlinearity. So the output is

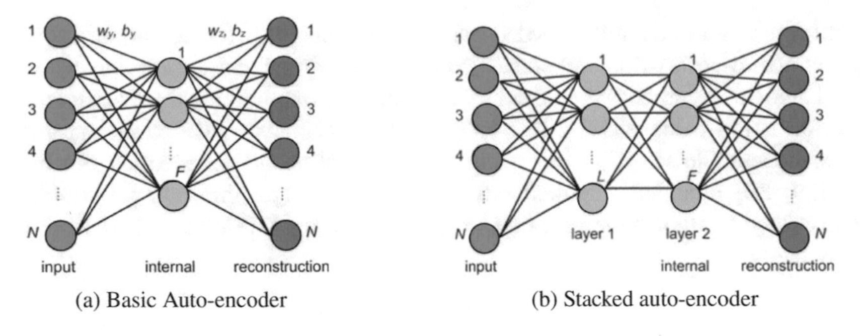

(a) Basic Auto-encoder (b) Stacked auto-encoder

Fig. 2.16 General SAE model. It learns a hidden feature from input

the decoded output based on the latent representation extracted by the auto-encoder. W_a, W_b are the weights between the hidden layer and the reconstruction layer. Each hidden unit extracts a particular feature and during reconstruction that feature is fed back into the decoder. We train the model such that the hidden representation maintains all the information about the input. For instance, if we use the hidden layer that is much smaller than the input layer, it means auto-encoder is going to compress the information, ignore the part of input that is not useful for reconstructing it and focuses on the part of the input that is more important to extract from it, for subsequent reconstruction. Therefore, it could be used to extract meaning full feature for classification.

$$C(x, z) = -\frac{1}{m}\sum_{k=1}^{d}\left[(x_k - z_k)^2 + (1 - x_{ik}) + \frac{\beta}{2}W^2\right] \tag{2.3}$$

Where x and z are the input and reconstructed data respectively.

2.5.2.2 Deep Belief Network

At the origin of the recent advances in DL, Deep Belief Network (DBN) is one of the major parts. It is considered as the origin of the unsupervised layer-wise training procedure. DBN is based on a lot of important concepts in training deep neural networks that are probabilistic in nature. DBN is a generative model that mixes undirected and directed interactions between the variables that constitute either the input or visible layer or all the hidden layers. As shown in Fig. 2.17 as an example. Here we have undirected connections in the start from input to hidden layer but the directed connection from hidden to hidden and hidden to output layer. In DBN, the top 2 layers always form an RBM. In Fig. 6.15, distribution over h^2 and h^3 is an RBM with undirected connections. While other layers are going to form Bayesian Network with directed interactions.

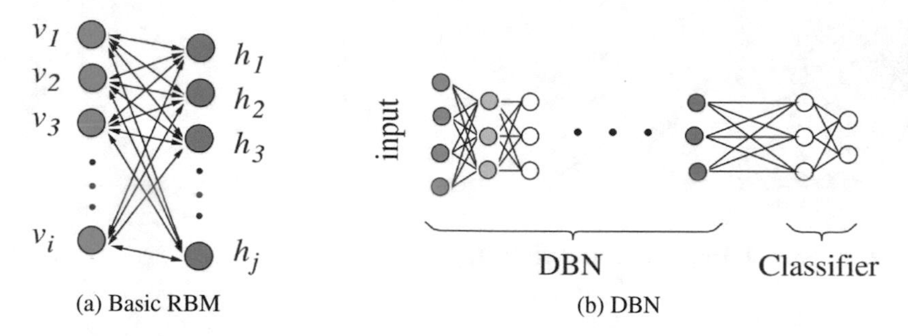

(a) Basic RBM
(b) DBN

Fig. 2.17 General DBN model

This is going to correspond to the probabilistic model associated with the logistic regression model. DBN is not a feed-forward network. Specifically, joint distribution over the input layer and three hidden layers are going to be prior over h_2 and h_3. Training a DBN is a hard problem. Good initialization can play a really crucial role in the quality of results. The idea of the initialization procedure is that in order to train, for instance, 3 hidden unit DBN, we take parameters of the first RBM as an input to the second RBM and so on. After this, there is a fine-tuning procedure that is not backpropagation. The algorithm for fine-tuning is known as the up-down algorithm.

2.5.2.3 Convolutional Neural Networks

Convolutional Neural networks or Conv nets or CNNs is an artificial neural network so far has been popularly used for analyzing images. Although image analysis is the most widespread use of CNNs, they can also be used for other data analysis such as classification problems. CNNs are designed from biologically driven models. Researchers found that human beings are perceiving the visual information in structured layers. Most generally we can take CNN as an artificial neural network that has some specialization for being able to detect patterns and make sense of them. This pattern detection is what makes CNN useful for image analysis. It is a trainable multi-layer architecture comprising of an input layer, a set of hidden convolutional layers, and an output layer as Fig. 2.18 presents. Generally, CNN consists of several feature extraction phases. Each phase comprises three layers, hidden Convolutional layer, pooling layer, and nonlinear layer. A typical CNN comprises of two or three such phases for deep feature extraction, and then one or more typical fully connected layers and then a classifier at the top, to classify the learned features. CNN can extract the deep representation but the main bottleneck in HSI classification is the limited label data as CNN requires a lot of training data to learn the parameters. Each phase is briefly explained in the following section.

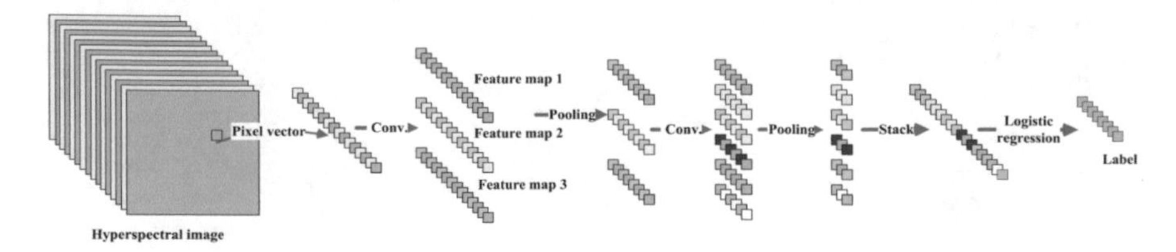

Fig. 2.18 General framework of deep CNN [37]

Convolutional Layer

Convolutional layer is a set of filters that are applied to a given input data. In case of HSI, the input to convolutional layer is a three-dimensional data with n two-dimensional features each of size $r \times c$. The output of the convolutional layer is also a three-dimensional data of size $r_l \times c_l \times l$. Where l is the size of features each of size $r_l \times c_l$. Convolutional comprises of filter banks which connects the input to the output.

Nonlinear Layer

Nonlinear layer is a CNN that comprises an activation function that takes the features generated as an output by convolutional layer and generates the activation map as an output. Activation function comprises element-wise operation so the size of input and output is the same.

Pooling Layer

It is a kind of nonlinear downsampling layer, responsible for decreasing the spatial dimensional of activation maps. Generally, they are used after convolutional and nonlinear layers to reduce the computational burden.

2.5.2.4 Sparse Coding Model

Sparse coding is a model in the context of unsupervised learning. In the context where we have training data that is not labeled, i.e., set of x vectors in the training set. It helps us to extract meaningful features from the unlabeled training set and allows us to leverage the accessibility of unlabeled data. A great number of sparse-based techniques have been proposed for HSI classification.

Any input x, seek the latent representation h that is sparse that means h consists of many zeros and only a few none zero elements. We also want the latent representation to contain meaning full information about x to be able to reconstruct the original input. The objective function for these conditions can be formulated as

$$\min_{D} \frac{1}{M} \sum_{k=1}^{M} \min_{h^{(k)}} \frac{1}{2} \left\| [(x^k - Dh^k) \right\|_2^2 + \beta \left\| h^{(k)} \right\|_1 \tag{2.4}$$

Where the first part of the equation represents the reconstruction error that we want to minimize. Matrix D refers to a dictionary matrix. The next term is sparsity penalty as we want latent representation to be sparse, for this we will penalize the l_1 norm. While β term controls to what extent we wish to get a good reconstruction error compared to achieving high sparsity. So that is the objective we want to optimize for each training example $x^{(t)}$.

References

1. Website, howpublished = https://en.wikipedia.org/wiki/electromagnetic_spectrum note = .
2. http://www.microimages.com/documentation/Tutorials/introrse.pdf
3. http://research.csiro.au/qi/spectroscopy-and-hyperspectral-imaging
4. Ghamisi P, Yokoya N, Li J, Liao W, Liu S, Plaza J, Rasti B, Plaza A (2017) Advances in hyperspectral image and signal processing: a comprehensive overview of the state of the art. IEEE Geosci Remote Sens Mag 5(4):37–78
5. National ecological observatory network, Battelle, Boulder, CO, USA. http://data.neonscience.org
6. Li J, Bioucas-Dias JM, Plaza A (2013) Spectral–spatial classification of hyperspectral data using loopy belief propagation and active learning. IEEE Trans Geosci Remote Sens 51(2):844–856
7. Camps-Valls G, Tuia D, Bruzzone L, Benediktsson JA (2014) Advances in hyperspectral image classification: Earth monitoring with statistical learning methods. IEEE Signal Process Mag 31(1):45–54
8. Benediktsson JA, Palmason JA, Sveinsson JR (2005) Classification of hyperspectral data from urban areas based on extended morphological profiles. IEEE Trans Geosci Remote Sens 43(3):480–491
9. Zhang L, Huang X, Huang B, Li P (2006) A pixel shape index coupled with spectral information for classification of high spatial resolution remotely sensed imagery. IEEE Trans Geosci Remote Sens 44(10):2950–2961
10. Huang X, Lu Q, Zhang L (2014) A multi-index learning approach for classification of high-resolution remotely sensed images over urban areas. ISPRS J Photogramm Remote Sens 90:36–48
11. Jia X, Kuo B-C, Crawford MM (2013) Feature mining for hyperspectral image classification. Proc IEEE 101(3):676–697
12. Zhang L, Zhang L, Tao D, Huang X (2012) On combining multiple features for hyperspectral remote sensing image classification. IEEE Trans Geosci Remote Sens 50(3):879–893
13. Zhang L, Zhang Q, Zhang L, Tao D, Huang X, Du B (2015) Ensemble manifold regularized sparse low-rank approximation for multiview feature embedding. Pattern Recognit 48(10):3102–3112
14. Melgani F, Bruzzone L (2004) Classification of hyperspectral remote sensing images with support vector machines. IEEE Trans Geosci Remote Sens 42(8):1778–1790
15. Li W, Tramel EW, Prasad S, Fowler JE (2014) Nearest regularized subspace for hyperspectral classification. IEEE Trans Geosci Remote Sens 52(1):477–489
16. Benediktsson JA, Pesaresi M, Amason K (2003) Classification and feature extraction for remote sensing images from urban areas based on morphological transformations. IEEE Trans Geosci Remote Sens 41(9):1940–1949
17. Bengio Y, Courville A, Vincent P (2013) Representation learning: a review and new perspectives. IEEE Trans Pattern Anal Mach Intell 35(8):1798–1828
18. Zhang L, Zhang L, Du B (2016) Deep learning for remote sensing data: a technical tutorial on the state of the art. IEEE Geosci Remote Sens Mag 4(2):22–40

19. Acito N, Diani M, Corsini G (2011) Signal-dependent noise modeling and model parameter estimation in hyperspectral images. IEEE Trans Geosci Remote Sens 49(8):2957–2971
20. Roger R, Arnold J (1996) Reliably estimating the noise in AVIRIS hyperspectral images. Int J Remote Sens 17(10):1951–1962
21. Gao L-R, Zhang B, Zhang X, Zhang W-J, Tong Q-X (2008) A new operational method for estimating noise in hyperspectral images. IEEE Geosci Remote Sens Lett 5(1):83–87
22. Amer A, Dubois E (2005) Fast and reliable structure-oriented video noise estimation. IEEE Trans Circuits Syst Video Technol 15(1):113–118
23. Ghazal M, Amer A, Ghrayeb A (2007) A real-time technique for spatio–temporal video noise estimation. IEEE Trans Circuits Syst Video Technol 17(12):1690–1699
24. Lee J, Hoppel K (1989) Noise modeling and estimation of remotely-sensed images. In: Geoscience and remote sensing symposium, 1989. IGARSS'89. 12th Canadian symposium on remote sensing, 1989 International, vol 2. IEEE, pp 1005–1008
25. Liu X, Tanaka M, Okutomi M (2012) Noise level estimation using weak textured patches of a single noisy image. In: 2012 19th IEEE international conference on image processing (ICIP). IEEE, pp 665–668
26. Sun W, Zhang L, Du B, Li W, Lai YM (2015) Band selection using improved sparse subspace clustering for hyperspectral imagery classification. IEEE J Sel Top Appl Earth Obs Remote Sens 8(6):2784–2797
27. Arzuaga-Cruz E, Jimenez-Rodriguez LO, Velez-Reyes M (2003) Unsupervised feature extraction and band subset selection techniques based on relative entropy criteria for hyperspectral data analysis. Proc SPIE 5093:462–473
28. Bajcsy P, Groves P (2004) Methodology for hyperspectral band selection. Photogramm Eng Remote Sens 70(7):793–802
29. Guo B, Gunn SR, Damper RI, Nelson JD (2006) Band selection for hyperspectral image classification using mutual information. IEEE Geosci Remote Sens Lett 3(4):522–526
30. Chang C-I, Wang S (2006) Constrained band selection for hyperspectral imagery. IEEE Trans Geosci Remote Sens 44(6):1575–1585
31. LeCun Y, Bengio Y, Hinton G (2015) Deep learning. Nature 521(7553):436
32. Mohamed A-R, Sainath TN, Dahl G, Ramabhadran B, Hinton GE, Picheny MA (2011) Deep belief networks using discriminative features for phone recognition. In: 2011 IEEE international conference on acoustics, speech and signal processing (ICASSP). IEEE, pp 5060–5063
33. Collobert R, Weston J (2008) A unified architecture for natural language processing: deep neural networks with multitask learning. In: Proceedings of the 25th international conference on machine learning. ACM, pp 160–167
34. Bengio Y, Lamblin P, Popovici D, Larochelle H (2007) Greedy layer-wise training of deep networks. In: Advances in neural information processing systems, pp 153–160
35. Krizhevsky A, Sutskever I, Hinton GE (2012) ImageNet classification with deep convolutional neural networks. In: Advances in neural information processing systems, pp 1097–1105
36. Serre T, Kreiman G, Kouh M, Cadieu C, Knoblich U, Poggio T (2007) A quantitative theory of immediate visual recognition. Prog Brain Res 165:33–56
37. Chen Y, Jiang H, Li C, Jia X, Ghamisi P (2016) Deep feature extraction and classification of hyperspectral images based on convolutional neural networks. IEEE Trans Geosci Remote Sens 54(10):6232–6251

Chapter 3
Unsupervised Hyperspectral Image Noise Reduction and Band Categorization

This chapter presents a thorough study and development of the algorithm for the first step toward HSI classification, i.e., noise/redundancy detection as shown in Fig. 3.1. A complete description of all the HSI classification phases is depicted in Chap. 1, Fig. 1.3. This phase aims at the detection of noise and redundancy for the classification of remote sensing hyperspectral images by addressing a number of issues.

As discussed earlier, the hyperspectral image consists of hundreds of spectral channels with a significant amount of redundancy and noise that not only makes the subsequent analysis of HSI, really challenging and difficult but also degrades the performance of subsequent classification algorithms. Over the most two recent decades, progression has been witnessed in hyperspectral sensors. In the area of remote sensing, analysis of the hyperspectral image is observed as one of the rapidly developing innovations. There are hundreds of thin adjoining spectral bands included in hyperspectral data. In the widespread area of applications, this kind of rich data becomes remarkably important. These applications include environmental surveillance, mineral detection, environmental monitoring, precision farming, environmental management, and urban planning. Although there is a substantial value of noise and redundancy involved in the hyperspectral images, that causes geometric and statistical properties that result in the problematic and reduces effectiveness for the successive Hyperspectral image data applications and analysis. This includes spectral unmixing, data display, data storage, data transmission, identification of the data, categorization of the data, and data processing. From the scientific and operational point of view, while creating its abilities and limitations it is an important step that it can assess the hyperspectral band properties. On the basis of data included in every band and band quality, an adaptive boundary band characterization has been suggested in this study for the first time. This makes bands to be available for all consequent applications. Bands with dissimilar possessions and features are required

L. Tao and A. Mughees, *Deep Learning for Hyperspectral Image Analysis and Classification*, Engineering Applications of Computational Methods 5,
https://doi.org/10.1007/978-981-33-4420-4_3

for different applications. From these classified bands on the basis of characteristics, it is possible to select the subsets of bands. As an example, the bands with dissimilar features are required for object detection, classification, denoising, noise estimation, and band selection. On the basis of the movement of boundary movement or boundary adjustment tri-factor criteria with the utilization of local structural regularity and self-similarity, the suggested approach, allows the classification of bands based on the level of information contained in each band. For the consequent applications, issues can also be solved with efficient band characterization. This comprises of Hughes phenomenon without making causing any issues to the original spectral meaning of a raw dataset of hyperspectral image. By implementing the proposed method to the hyperspectral image classification and noise estimation, the application of this framework is demonstrated. On the basis of the segmented regions obtained from the adaptive boundary adjustment and movement execution of noise estimation is done. On the basis of the new suggested segmented regions and noise model, the classification and noise estimation provide comparable output. The hyperspectral images originated from the hyperspectral sensors have reliability and quality that is sensitive to noise [1]. The attained data consists of a substantial quantity of instrument noise, bands that are noisy, and momentary externalities that are linked with regular water absorption features, even due to the progression in the hyperspectral images. Image with high Signal-to-Noise Ratio (SNR) is needed for many hyperspectral images applications which includes, classification [2], signal-subspace identification [3], spectral unmixing, information extraction, analysis, scene interpretation, etc. Not only the precision of consequent applications is affected by this excess noise, but also for subsequent applications [2]. Therefore, detection of noise is considered an imperative job and altogether influences further HSI examination. For instance, in airborne noticeable/infrared-imaging spectrometer (AVIRIS) and intelligent optics framework imaging spectrometer-sensor (ROSIS) datasets [4], a considerable lot of the phantom groups have a high SNR; in any case, as per analyst experience, a numerous number of groups (up to 20%) are very loud. A few applications dispose of the boisterous groups, despite the fact that this may bring about loss of valuable information [5, 6]. The versatile limit based band arrangement (A3BC) calculation permits estimation of the data/commotion level and recoups helpful data from uproarious groups so as to make it accessible for further examination and application.

There are two main categories for hyperspectral images noise [7]. One is fixed pattern noise and the other one is random noise. Random noise is stochastic in nature, so it could not be eradicated easily. There are specific noise models for the processing of hyperspectral image algorithms, due to which their performance is remarkably degraded if these models did not address the characteristics of noise properly. The additive model is mostly used as a random noise model in hyperspectral images. The majority of literature addressing the processing of hyperspectral does not cater for atmospheric effects. Other noise models did not reduce the effect of these atmospheric effects. Before the rays enter the detector, for example, water vapor absorption noise contamination is encountered by the hyperspectral images process. During the literature review, we have observed the noise as a composition of detector noise and scene noise. Scene noise is caused by interactions between radiance and the

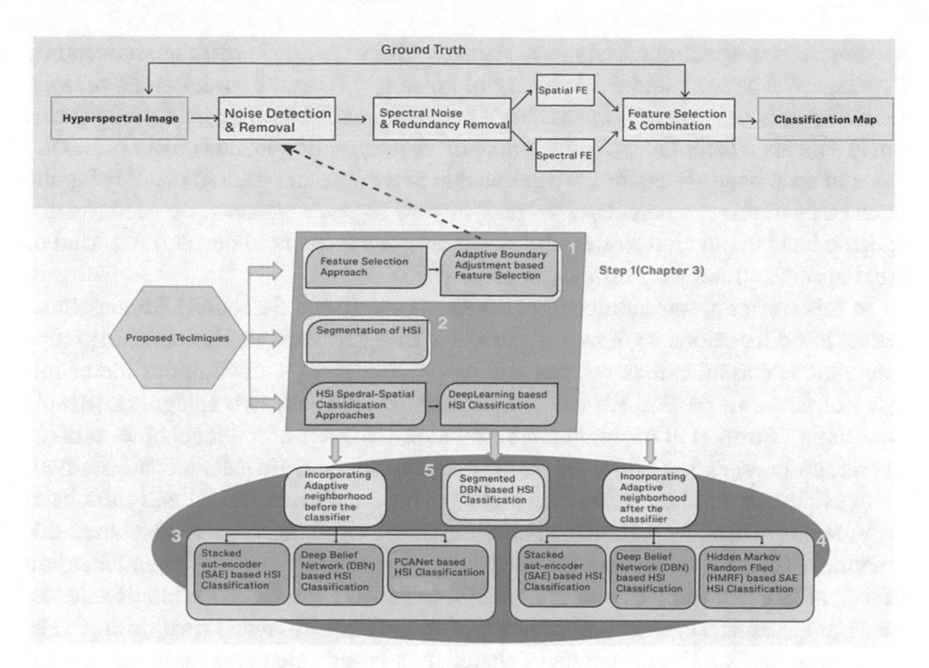

Fig. 3.1 First phase aligned with the framework toward the classification of hyperspectral remote sensing images

atmosphere and is mainly composed of path-scattered effects and absorption effects. Scene noise is significant at certain wavelengths, whereas sensor noise is mainly derived from thermal energy and shot noise. There are signal-dependent elements in both detector noise and scene noise.

Recently there are three main categories in which noise can be classified. These are filter-based, block-based, and it could be a combination of both. Images in the block-based approach are first divided into a number of dense blocks, after this block is sorted out with minimal texture and structure. Based upon these blocks noise variance is estimated. Usually, the structure measurement in these approaches is made by spatiotemporal homogeneity, variance, spatial homogeneity, and gradient covariance matrices.

Recent methods that are being used for noise estimation are segmentation based and pixel growing. To define the same object segments for estimation of covariance-matrix based on extremely non-uniform optically shallow environment for image-based segmentation is used by Sagar et al. Likewise, the method described in ref 21, that is, kernel-based method is an edition of pixel growing approach which is basically developed by Dekker and patters. Similar methods were adopted in remote sensing scenarios for studies of optically shallow water.

There are a number of disadvantages we have to face whenever we are using these approaches. Firstly during processing, we have to treat special and spectral dimensional equally but in actual hyperspectral images, spatial correlation is much lesser as compared to spectral correlations. Along with this, mixed noises are induced by hyperspectral sensors and the intensity of noise is different between most sensors. Gaussian noise having uniform variance between bands could be handled by previously described techniques. Another unrealistic assumption made is that noise is the same in each band. Therefore, we are unable to achieve the desired result using the actual data of these approaches. Keeping in view these drawbacks, we need to use a specific band estimation strategy for noise. So, our focus is to develop this kind of band-specific strategy for estimation of noise.

In this research, we additionally considered the (Band Selection) BS classification of low dimensionality, however, permitted the adaptability of isolating all groups into various classifications without wiping out any groups, making our methodology not the same as BS. We can categorize BS work into two categories [8]: (1) maximum information or minimum correlation (MIMC)-based techniques and (2) maximum interband separability (MIS)-based techniques. MIMC techniques typically use intraband-correlation and cluster criterion. Cluster model and Intra band correlation are used by MIMC methods. The intraband-connection model based calculation assembles reasonable subsets of groups by amplifying the general measure of data utilizing entropy like measurements. In 24, 25 MIS-based calculations select the appropriate arrangement of groups having the least intraband relationships. For instance, Refs. 26, 27 incorporated a shared data-based calculation and a compelled band-choice calculation dependent on obliged vitality minimization, individually.

In this investigation, we proposed a ghostly spatial, versatile limit-based band arrangement (AB3C) system, where contrasts in clamor power between adjoining groups and spatial attributes are both considered. AB3C could likewise be considered as an underlying advance for hyperspectral images commotion estimation, denoising, BS, characterization among others. In this examination, hyperspectral images' commotion estimation and arrangement are likewise executed as an application dependent on AB3C. In hyperspectral images, neighborhood force varieties are because of reflectivity, illumination from the picture itself, or from noise. Utilizing picture fluctuation as an estimation of commotion level is probably going to result in overestimation; in this way, we sectioned the hyperspectral images into bunches and doled out certainty scores to the groups dependent on the degree to which the variety was brought about by clamor. Rather than utilizing the credulous mean of the middle to gauge commotion level, we fit a clamor level capacity with the majority of the groups weighted by certainty scores. Since clamor levels differ with groups, we had the option to gauge commotion levels on each band, individually. A noise model is proposed for the estimation of noise, which combines different factors like instrument-based SNR estimation, variation of environment that affect the scene explicitly, interference of air and water, and diffused refractions of daylight and direct sky. The images are divided into three bands based on the cluster in the proposed philosophy (28–30). Strategies based on clusters are surprisingly delicate to noise, as certain groups may comprise totally of boisterous information that can be evaluated utilizing limit alteration-based criteria autonomously. The clamor level of each band would then be able to be determined dependent on different limit change factors. The primary commitments of the proposed system are outlined as follows: The proposed methodology is based on cluster-based segmentation of three-band images [9–11]. Cluster-based methods are extremely sensitive to noise, as some clusters may consist entirely of noisy data that can be estimated using boundary adjustment based criteria independently. The noise level of each band can then be calculated based on various boundary adjustment factors.

The main contributions of the proposed framework are summarized as follows:

- (1) The first most criteria for the spatially adaptive method for boundary adjusted hyperspectral images is AB^3C;
  (2) The property used to capture the information that each band carry is edge movement, Due to which with the aid of band properties we are able to get a new framework;
  (3) Variation of information among the different bands and spatial information differences between the bands is being considered in this model; and (4) Estimation process of noise based on band spatial adaptive is automated.
- • The level of noise for all the bands is computed according to the constructed tri-weight factor estimation of noise is done base on the clusters created based on segmented regions.
- • To cover both noises scene noise and sensor noise based on sensor and atmosphere target strategy a model for noise is suggested.

The proposed framework has two main advantages:

- • According to the level of information, bands are characterized and identified and useable to eradicate bands with only noise and bands having little or no data based upon the application.
- • Applications where we can use it effectively used are, denoising, noise estimation, classification BS, and many others.

The performance of the proposed algorithm was validated on both a synthetic dataset (for noise estimation) to evaluate its performance in a controlled environment with band-specific Gaussian noise and mixed noises (Gaussian, Poisson, and spike noise) and real HSIs.

The organization of the remainder of this chapter is as follows: Sect. 3.2 reviews the related algorithms, provides support for the proposed methodology, and defines, in detail, each component of the proposed framework and Sect. 3.3 describes the extensive experimentation used to examine the capabilities of the proposed framework followed by the summary.

3.1 Proposed Methodology

In this section, we will briefly discuss the suggested methodology for the noisy band detection based on boundary adjustment based unsupervised framework. Figure 3.2 depicts the overview of the flowchart and is being concluded in Algorithms 1 and 2. In this framework, the most important part of this framework is to evaluate the

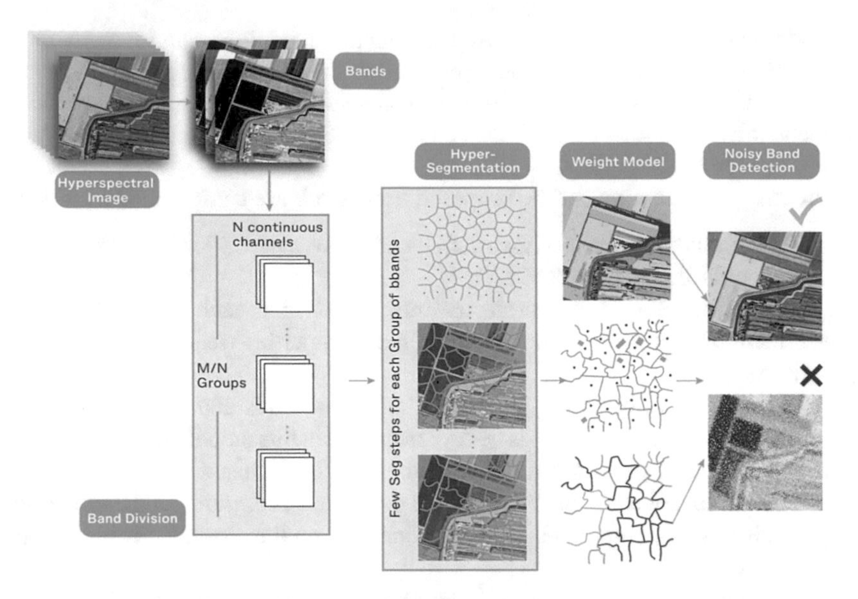

Fig. 3.2 Flowchart of the boundary adjustment based noisy band selection framework

noisy band using weight-based criterion. To estimate the noisy channel, the data comprising M channels is divided into $\frac{M}{N}$ G-channel groups, each divided data is analyzed independently that resulted in a band-noise factor $W \in [0, 1]$ to exploit noise level.

The data comprises pixels $p_{i,j}$ (with $i \in [1, width]$, $j \in [1, height]$) and is divided into pixel clusters ("cells"), $C_i \in \mathbf{C}$, with members being $p \in C_i$. Each pixel property is observed as actual valued, multidimensional vector, $x_p \in [0, 1]^N$, with each dimensional being one band's intensity value of this pixel.[1] Initially, using a hexagonal region, R clusters are placed on an image.

So, every cluster would have an average special spectral contextual expression.

$$SSC(C_i) = \frac{1}{|C_i|} \sum_{p \in C_i} x_p \tag{3.1}$$

For each boundary pixel, we have a straightness factor

$$\tilde{n}_p(C_i) = |\{p' \notin C_i, dist(p, p') \leq r\}| \tag{3.2}$$

To show the number of pixels on connected disk, there is a straightness factor which is linked to different clusters with a constant factor r (radius of the neighborhood disk).

[1]The intensity value of HSI data is linearly normalized into a [0, 1] range using the band-wise maximum and minimum intensities.

For boundary pixels, we denote the ownership cluster as $O(p)$ s.t. $p \in O(p) \forall p$ and define neighboring clusters

$$N(p) = \{O(p') | dist(p, p') \leq 1\} \tag{3.3}$$

Algorithm 1: Adaptive cluster-edge adjustment

Input : HSI Bands $\mathcal{B}$, with pixels, p, and intensity vectors, x_p, and initial clusters $\{P_k\}_{(k=1..K)}, C_{k,0} = P_k$.
Output: boundary adjustment-based final clusters/segments

```
1  bBoundaryAdjustment ← true;
2  i ←0;
3  while bBoundaryAdjustment do
4      i ← (i + 1);
5      bBoundaryAdjustment ← false;
6      for each (p_b, C_k,i) (boundary pixel p_b, adjacent cluster C_k,i pair) do
7          Compute the class centroid, SSC, as defined in (1);
8          Compute p_b's local gradient, grad(p_b), as defined in (4);
9          Compute p_b's straightness factor as defined in(2);
10         Using the above values, compute the cluster- energy function, E_rET(p_b, C_k,i),
            defined in (5);
11         for All Pixels p do
12             compare E_rET(p_b, C_k,i) for each neighbouring cluster;
13             Re-assign it to the cluster with the largest E_rET;
14             Find k, such that E_rET(p, C_k,i) is the maximum;
15             put p into C_k,i+1;
16             if C_reassigned ≠ C_current then
17                 bBoundaryAdjustment ← true;
18             end
19         end
20     end
21 end
```

to be the set of owning clusters of the adjacent pixels. We further define the pixel position gradient using a four-way kernel method:

$$grad(p_{i,j}) = \frac{1}{4} \|x_{p_{i,j}} - x_{p_{i+1,j}}\| + \|x_{p_{i,j}} - x_{p_{i,j+1}}\| + \|x_{p_{i,j}} - x_{p_{i-1,j}}\| + \|x_{p_{i,j}} - x_{p_{i,j-1}}\| \tag{3.4}$$

Finally, using an approach similar to that of [9], we define the clustering-energy function between boundary pixel and adjacent cluster as:

$$E_{rET}(p, C_i) = \sqrt{\|x_p - SSC(C_i)\| + \lambda \times \tilde{n}_p(C_i) \times grad(p)} \tag{3.5}$$

for the robust Estimation (rET) algorithm, where λ shows a constant monitoring of the significance of the edge-straightness portion. By considering the gradient, the algorithm is more likely to adjust a boundary to an actual image-segmentation boundary, which is an important characteristic of a non-noisy image.

At every reiteration, every reiteration pixel computes the vitality against every single neighboring cluster and characterizes the new cluster like the one with the biggest vitality. If there is any difference between two clusters then the pixel is expelled from its previous cluster and added to the new cluster. The calculation ends when no pixels get the new value after one cycle, or the quantity of new assignments in the emphasis is not exactly a limit steady.

The Band-Noise Factor (BNF) depicts a mixture of various sub-factors, each normalized in the range [0, 1] computed dependent on the boundary being adjusted set up by results from the above calculation. We have defined each factor segmentation in the next paragraph.

3.1.1 Preprocessing Toward Initial Segmentation

Each pixel is defined as an N-dimensional vector if we are provided with a group of N-channel. Hyperspectral image segmentation is found by running a prescribed algorithm. To be more specific, we would have a set of clusters, $\mathbf{C} = (C_1, \ldots, C_R)$, and a boundary definition, $\mathbf{E} = \{p|N(p) > 1\}$. The factors that would contribute to noisy band classification and detection are as follow.

3.1.2 Cluster-Size Factor

A well-adjusted criterion is choosing for the initial setup of the adjusted boundary of a cluster for a noisy image. Usually, each cluster has an identical mean intensity that results in lower boundary differentials. In this, we would be able to complete the adjustment for various regions. There would be a minimal difference in cluster size being compared to the initial setup.

Given the size of clusters, $|C_i|$, we determine its standard deviation (variance of size) as

$$s_C = \sqrt{\frac{1}{R}\sum_i \left(|C_i| - \overline{|C_i|}\right)^2} \tag{3.6}$$

where the average is

$$\overline{|C_i|} = \frac{height \times width}{R} \tag{3.7}$$

The factor is compared to a constant threshold and mapped to [0, 1]:

$$W_{size} = \min\left(1, \frac{s_C}{\lambda_{size} \times \overline{|C_i|}}\right) \tag{3.8}$$

3.1.3 Cluster-Shift Factor

Similarly, it may be the reason for not achieving adjustment that means the groups of segmentation are not varying sufficiently from its initial value. Whereas, if we have a cluster of informative channel, the position of the boundary will meaningfully change the positions to achieve the original boundary. A little movement of cluster focuses, in this manner, suggests a higher likelihood of a noisy channel. We characterize the shift factor as follows:

$$shift(C_i) = dist(center(C_i), center(C_{0,i})) \tag{3.9}$$

where the center of cluster C_i is defined as its center of mass with $C_{0,i}$ being the initial cluster from the hexagonal Voronoi region before adjustment.

The weight factor is defined as

$$W_{shift} = \min\left(1, \frac{\overline{shift(C_i)}}{\lambda_{shift} \times D}\right) \tag{3.10}$$

where the shift is compared with the initial cluster diameter, D, which can be approximately calculated as

$$D = \sqrt{\frac{H \times W}{R} \times \frac{4}{\pi}} \tag{3.11}$$

3.1.4 Cluster Spatial-Spectral Contextual Difference Factor

Two different surface items are represented by two contiguous groups in hyperspectral images when a real value boundary fits the cluster at the point when and these would have specifically normal SSC data conditioned an unremarkable difference channel group. Interestingly, in a noisy channel gathering, all groups have around a similar normal SSC data, paying little attention to the modification. Subsequently, the bigger contrast in normal SSC infers a bigger measure of data on the boundary adjustment.

For each boundary edge pixel, $e \in \mathbf{E}$, with adjacent cluster C_a and C_b, we define the SSC difference as

$$cd(e) = \|SSC(C_a) - SSC(C_b)\|_1 \tag{3.12}$$

and the weight of the SSC difference as

$$W_{ssc} = \min\left(1, \frac{\overline{cd(e)}}{\lambda_{ssc}}\right) \tag{3.13}$$

3.1.5 Band-Noise Factor (BNF)

The BNF for a specific channel group is defined as the product of all of the above weight sub-factors (with averaging over the image) as

$$W = W_{size} \times W_{shift} \times W_{ssc} \tag{3.14}$$

As the range of values for weight sub-factors altogether are in [0, 1], we also have $W \in [0, 1]$. This factor represents the extent of useful data present in the specified channel group, with lesser weights representing the existence of more noise. An extremely low BNF, e.g., lower than a threshold, W_{min}, represents a noise-channel group.

Algorithm 2: AB^3C

Input : HSI $\mathcal{I}$ with M channels
Output: categorized band based on energy/noise

```
1 Divide M channels into M/N G-channel groups;
2 for i ← 1 to G do
3     Apply Algorithm 1 on each channel group;
4     Compute weight subfactor based on cluster-size variance, as defined in (6)-(8) ;
5     Compute weight subfactor based on cluster-shift variance, as defined in (9)-(11) ;
6     Compute weight subfactor based on SSC difference between edges, as defined in (12)
       and (13) ;
7     Compute BNF, as defined in (14);
8     Classify the bands based on the BNF, as defined in (14);
9 end
```

3.1.6 The HSI Process and Noise Model

The HSI noise procedure is delineated as (1) rays received from the sun to ground are reflected by the scene, (2) the reflected rays are sent to the climate that reaches the sensor, and (3) the image senor translates the radiance energy into signals and, lastly, to images. Reference [12] gives a magnificent representation of the imaging procedure as appeared in Fig. 3.3. In most of the previous literature, noise is displayed as an added element and free of the signal; in any case, noise figure is not that simple it usually has both signal-dependent and signal-free segments. From an image formation point of view, the last hyperspectral images are compiled by different noise sources. We displayed these noises as being made out of scene noise, emerging from the response of rays and the scene, and sensor noise, emerging from the change from rays to electronic sign in the sensor. Both noise sources have signal-reliant and signal-free parts. The perfect signal reaching the sensor is an unadulterated ray reflected by the object, whereas in reality the air communicates with the ray between the object and the sensor and is generally categorized by the dispersing and air retention

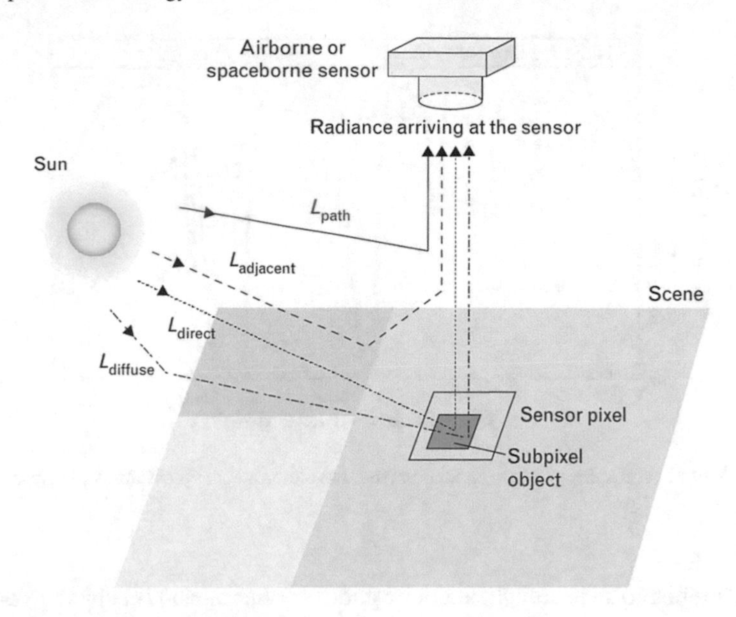

Fig. 3.3 Imaging system. Illumination sources and their path from the source to the surface of the scene and then into the sensor [12]

impacts. The atoms and particles in the air disperse the ray, forming path-scattered radiance. The all-out path-scattered rays landing at the sensor comprises of radiance comprising of disperse ray legitimately from the sun and close by surface-reflected radiance as narrated in detail in [12]. Scattering for the most part happens in the noticeable and close infrared wavelengths. Path-scattered radiance can be displayed as a constant offset plus zero-mean Gaussian noise. We displayed the perfect radiance entering the sensor as R and path-scattered radiance as $R_s = r_{ps} + n_{ps}$, where r_{ps} is a constant and n_{ps} depicts Gaussian noise.

Certain gases retain the vitality. Explicitly close to 1.4 and 1.9 μm, water vapor and carbon dioxide generally ingest all the vitality, resulting in junk groups in HSIs. The environmental transmittance is depicted in Fig. 3.4. We demonstrated the ingestion impact as the changes of photons due to their free movement through the air. The ingestion impact can be exhibited as decreasing the energy with a scaling term, a, controlled by transmittance, in addition to a zero-mean retention noise, n_a. As a result of discretization, the difference of n_a is relative to $\sqrt{R}$ with respect to central limit hypothesis. A genuine ray entering the senor is portrayed as

$$\hat{R} = aR + r_{ps} + n_a + n_{ps}. \tag{3.15}$$

The sensor noise sources are like shading Charge-Coupled Device (CCD) cameras. Following the noise model for CCD sensors in [14], we proposed a commotion model for hyperspectral images sensors. This incorporates thermal noise brought about by dull flows, which can be demonstrated as constant offset plus zero-mean Gaussian

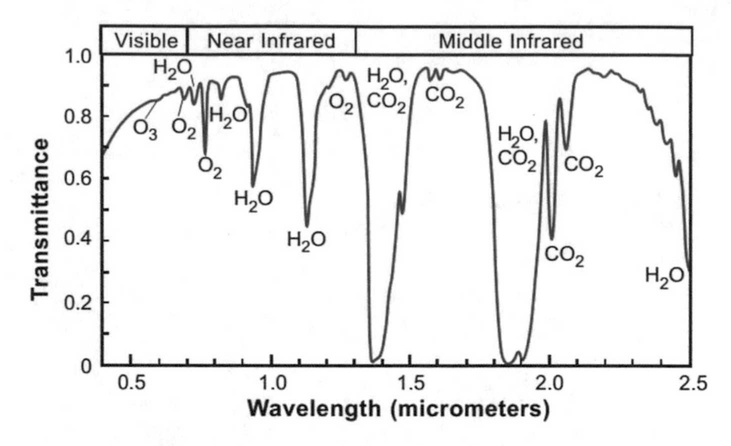

Fig. 3.4 Plot of atmospheric transmittance versus wavelength for typical atmospheric conditions [13]

noise. The shot noise is brought about by photon counting and is subject to radiance strength. Taking into account other noise sources, for example, quantization noise, enhancer noise, D/A and A/D noise, and added substance and sign free noise, n_s depicts shot noise, and we joined all sign autonomous noise sources into one term n_i with steady fluctuation. It results in

$$I = bL + c + n_s + n_i \tag{3.16}$$

where I is the perceived image intensity, L is the radiance of the object, n_s is the shot noise, and b is a scaling factor. In this literature, L is a random variable relying on $\hat{R}$. Joining (3.15) and (3.16), results

$$I = \alpha R + \beta + n_d + n_i, \tag{3.17}$$

where α is the scaling factor joining the absorption effect and photon-observed rate, β is the constant offset, n_d is the joined signal-dependent noise sources, and n_i is the joined signal-independent noise sources ($E(n_s) = 0$, $Var(n_s) = R\sigma_d^2$, $E(n_i) = 0$, and $Var(n_i) = \sigma_i^2$) and has constant variance.

3.1.7 Noise Estimation

Using the noise model described in Sect. 3.1.6, we follow the model to estimate the noise levels in hyperspectral data. Firstly, noise is depicted as

$$\tau(I) = \sqrt{E[(I_N - I)^2]} \tag{3.18}$$

where I_N is the observed data and $I = E(I_N)$. Precisely, this literature concludes

$$\tau(I) = \sqrt{I\sigma_s^2 + \sigma_i^2}. \tag{3.19}$$

Key decisive factors can be attained from the segmented outcome of the HSI data. These key factors include, $\{I_i, \sigma_i\}$, where I_i is the mean intensity and σ_i is the variance of that particular segmented region. These resulted key factors are then utilized to approximate the criteria for the level of noise in that particular segment. Bayesian Maximum a Posteriori (MAP) technique is explored to opt a lower rapper to the decisive factors. A lower rapper shows that the fitted function should have a high probability of being smaller than the approximate values and at the same time having a relatively high probability of being very close to the decisive factors. Mathematically, provided a specific noise-level indicator function, the responsibility of a likelihood function is to predict the HIS image intensity and variations in noise:

$$L(\tau(I)) = P(\{I_n, \hat{\sigma}_n\}|\tau(I)) \propto \prod_n \Phi\left(\frac{\sqrt{k_n}(\hat{\sigma}_n - \tau(I_n))}{\hat{\sigma}_n}\right) exp\left\{-\frac{(\tau(I_n) - \hat{\sigma}_n)^2}{2s^2}\right\} \tag{3.20}$$

where $\Phi(.)$ is the accumulative dissemination function of a typical normal dissemination, k_n is the sum of pixels in that particular segment, and s is the regularization factor utilized to govern the closeness of the key decisive factors to the function. Furthermore, the values of factors σ_s and σ_i are attained by resolving the conventional optimization:

$$\{\sigma_s^*, \sigma_i^*\} = \arg\min_{\{\sigma_s, \sigma_i\}} \sum_n \left[-\log \Phi\left(\frac{\sqrt{k_n}(\hat{\sigma}_n - \tau(I_n))}{\hat{\sigma}_n}\right) + \frac{(\tau(I_n) - \hat{\sigma}_n)^2}{2s^2}\right]. \tag{3.21}$$

The noise-approximation procedure is established as follows:

Algorithm 3: Estimating noise level

Input : Image I, segmentation $\{P_k\}_{k=1...K}$
Output: Noise level $\{\sigma_d, \sigma_i\}$
1 **for** *Each segment* P_k **do**
2 Calculate the pair (I_k, σ_k), where I_k represents the mean of all pixels and σ_k is the variance.
3 **end**
4 Define the noise-level function as $\tau(I) = \sqrt{I\sigma_d^2 + \sigma_i^2}$.
5 Define the likelihood function $L(\tau(I)) = P(\{I_n, \hat{\sigma}_n\}|\tau(I))$, which is the probability of seeing the samples, given a particular noise-level function.
6 Calculate $L(\tau(I))$ for all samples $\{(I_k, \sigma_k)\}$.
7 Find the parameters $\{\sigma_d, \sigma_i\}$ that maximize $L(\tau(I))$.

At a conceptual level, the proposed approach consists of the following main steps:

1. Group the initial HS M-channel image into $\frac{M}{N}$ N-channel groups.
2. Place K initial seeds for k classes into the HSI.
3. Divide the HSI into initial k hexagonal segments/clusters.

4. Iterate over the following basic steps until no further evolution is possible, i.e., no pixel transfers into a different segment.
    - Compute the class centroid of each segment.
    - Compute the gradient of each boundary pixel.
    - Compute the straightness factor for the boundary.
    - Compute the similarity measurement, i.e., attraction/repulsion forces between each class centroid and edge.
    - Evolve the boundary based on the similarity measurement.
5. Given the hyper-segmentation, compute all weight factors accordingly.
    - Compute weight subfactor based on cluster-size variance.
    - Compute weight subfactor based on cluster shift relative to the initial position.
    - Compute weight subfactor based on the SSC difference between edges.
6. Combine the sub-factors into the BNF.
7. Classify the bands based on the factor.

3.2 Experimental Results

The viability of the suggested AB3C structure is assessed in this segment by applying to specific hyperspectral images dataset, which is capable of cooperating with different quality of channels along with the channels that contain only noise. Their choice is upheld by the insights that these datasets have been generally utilized over the previous decade to approve hyperspectral images' investigation calculations on account of their multifaceted nature and assorted variety, and they are as of now used as benchmarks standards to approve novel calculations. For the estimation of noise, the strategy is applied to synthetic data information and genuine dataset.

3.2.1 HSI Datasets

In the experimental phase, we have used three different kinds of hyperspectral datasets being collected using AVIRIS sensors, Indian Pine (Fig. 3.5), Salinas Valley (Fig. 3.6), and Moffett Field (Fig. 3.7) captured by AVIRIS sensor is utilized. In-depth details of these challenging dataset are discussed in Sect. 2.3.

3.2.2 Synthetic Dataset

Dataset from [15] is used as a source of the synthetic dataset. The number of analysts already used this synthetic dataset for performance and evaluation of their proposed

Fig. 3.5 Indian Pines dataset and its ground truth boundaries

Fig. 3.6 Salinas dataset and its GT boundaries

algorithms [16]. The initial informational collection used in this set is of size 100 × 100, containing 221 groups each. Estimation of noise becomes difficult due to low resolution and commotion like surface.

3.2.3 Experiments on Real HS Data

This section details the credibility of our suggested A^3BC framework being applied on the datasets collected with the help of the AVIRIS sensor (Fig. 3.7).

Fig. 3.7 Moffett Field dataset

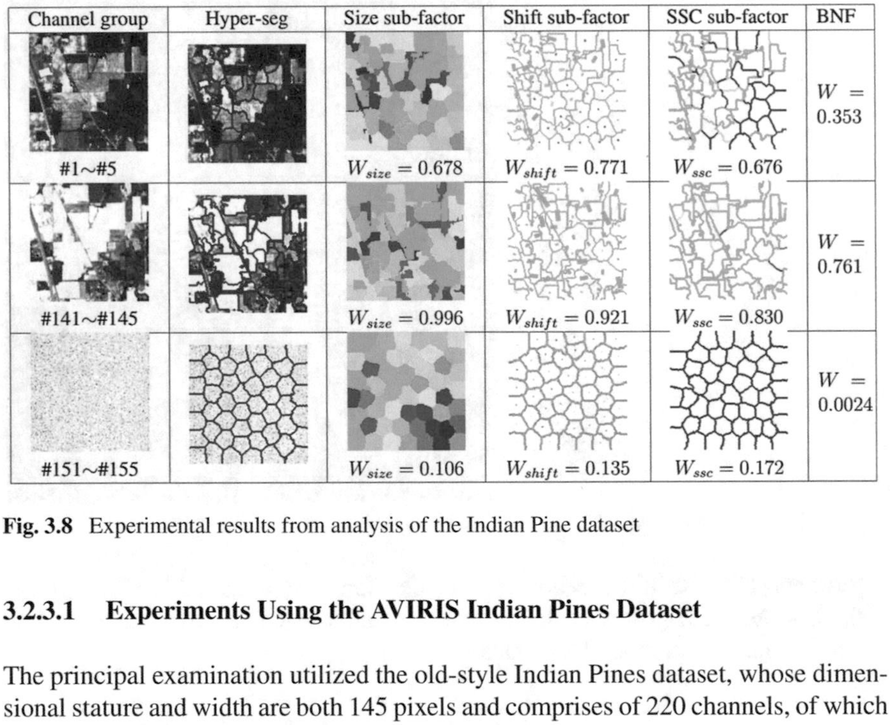

Channel group	Hyper-seg	Size sub-factor	Shift sub-factor	SSC sub-factor	BNF
#1∼#5		$W_{size} = 0.678$	$W_{shift} = 0.771$	$W_{ssc} = 0.676$	$W = 0.353$
#141∼#145		$W_{size} = 0.996$	$W_{shift} = 0.921$	$W_{ssc} = 0.830$	$W = 0.761$
#151∼#155		$W_{size} = 0.106$	$W_{shift} = 0.135$	$W_{ssc} = 0.172$	$W = 0.0024$

Fig. 3.8 Experimental results from analysis of the Indian Pine dataset

3.2.3.1 Experiments Using the AVIRIS Indian Pines Dataset

The principal examination utilized the old-style Indian Pines dataset, whose dimensional stature and width are both 145 pixels and comprises of 220 channels, of which 20 are proclaimed as noisy. The channels are assembled by five adjoining channels ($N = 5$) to be assessed by our framework.

The investigation of three normal direct channel groups is depicted in Fig. 3.8.[2] Channels from 141 to 145 are a normal channel, with edges similar to the real object edges, while channel group 1–5 are somewhat noisy. Channels 151–154 are

[2]For the cluster size representation, a bigger group is appeared in green, while a littler cluster appears in red. Both are representing non-noisy channels.

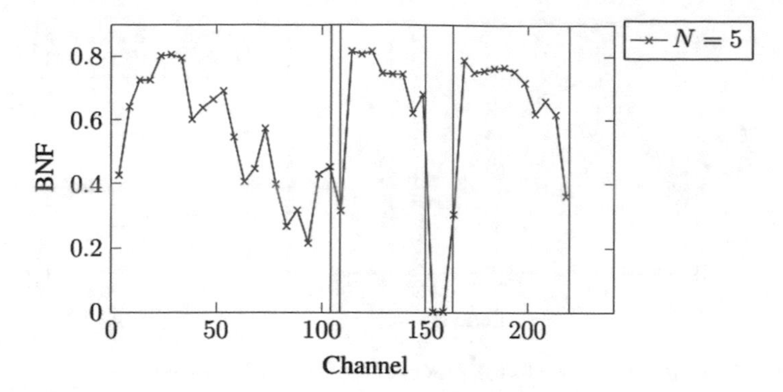

Fig. 3.9 BNF distribution over channel groups for Indian Pine dataset

unadulterated noisy, and the limit stayed in a hexagonal shape. The band-noise factor for all channels is depicted in Fig. 3.9. Authoritatively announced boisterous channels (water-engrossing wavelengths) are set apart by red lines. As can be observed in the differences, the proposed technique is equipped for recognizing completely noisy channel gatherings, while additionally doling out lower variables to partly noisy channels. The channels 151–165 are detected as completely noisy, while the channels 81–95, 106–110, and 161–166 are considered partly noisy.[3]

3.2.3.2 Experiment Using the Salinas Dataset

A subsequent examination was attempted utilizing the Salinas dataset. This dataset has 224 channels, with 20 announced as noisy. The measurements are 512×217 pixel.

Like the past investigation, the BNF and examination of all channels are depicted in Figs. 3.10 and 3.11, separately. Formally announced noisy channels (water-engrossing wavelengths) are set apart by red lines. As can be seen, the noisy bands by and large had BNF <0.2, while fully noisy bands (111–116 and 156–166) had BNF <0.01.[4] The variance in the mean BNF of noisy and regular-channels are depicted in Fig. 3.12.

Some channels with partly noisy channels were related to a lower BNF, yet none near zero. Because of the thin data bandwidth of the noise close to channel 110, we

---

For the cluster movement representation, a bigger group move is appeared in green with a more extended run, while a littler cluster move appears in red with a shorter run. A bigger move (longer run) is characteristic of a non-noisy channel.

For the boundary wise contrast delineation, a bigger distinction in group SSC is appeared by green edges, while a littler contrast is appeared by red edges. A bigger distinction is demonstrative of a non-noisy image.

[3]In the event that we use $W_{min} = 0.1$ as a limit, the following channels will be measured as noisy: 156–160 and 160–165.

[4]If $W_{min} = 0.1$ is utilized as the upper limit, following channels are marked as noisy:
$N = 5$: 106–110, 111–115, 156–160, 160–165, 166–170.
$N = 6$: 106–111, 109–114, 112–117, 151–156, 154–159, 157–162, 160–165, 163–168.

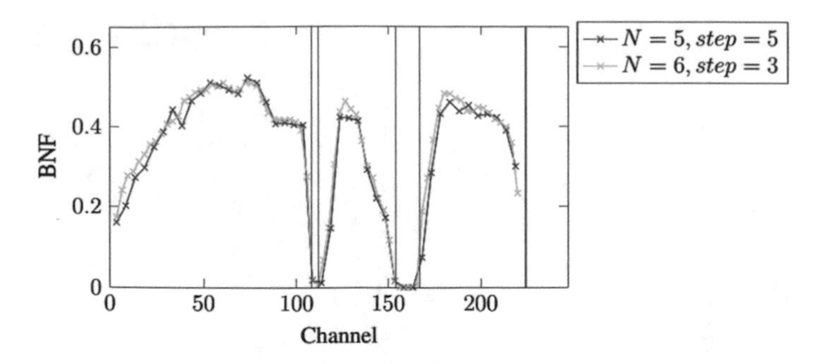

Fig. 3.10 BNF distribution over channel groups of the Salinas dataset

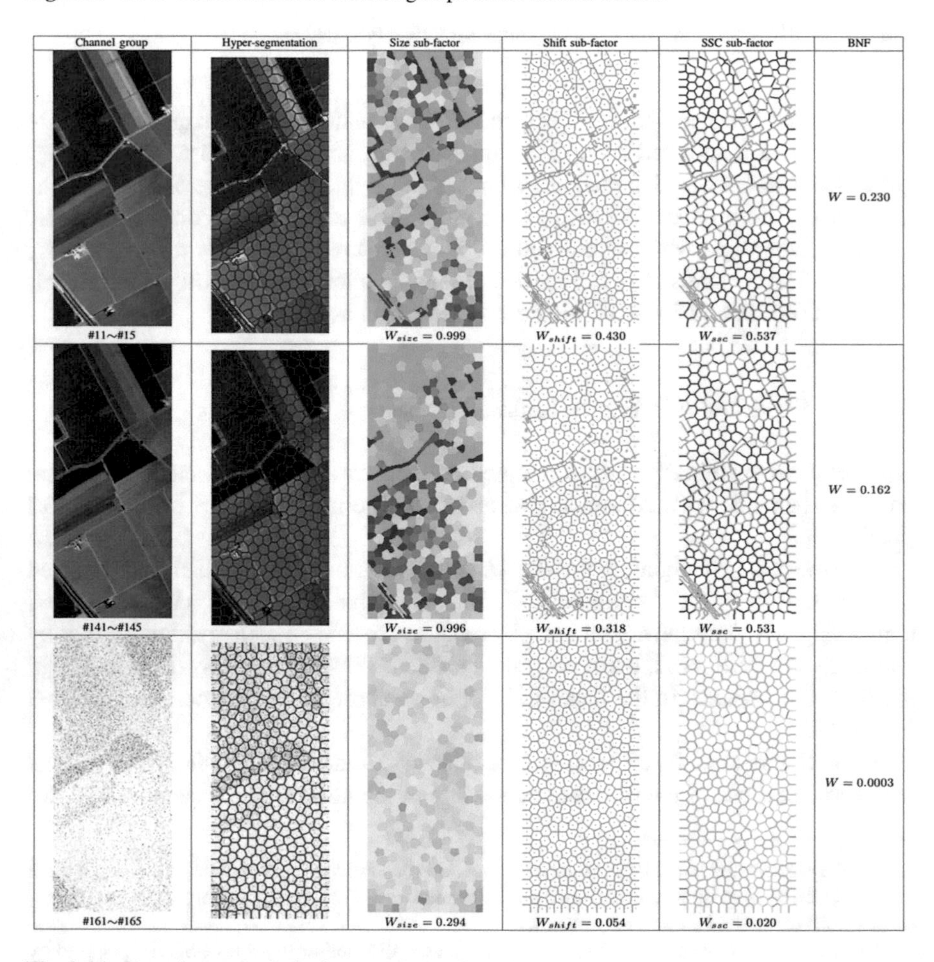

Fig. 3.11 Experimental results from analysis of the Salinas dataset

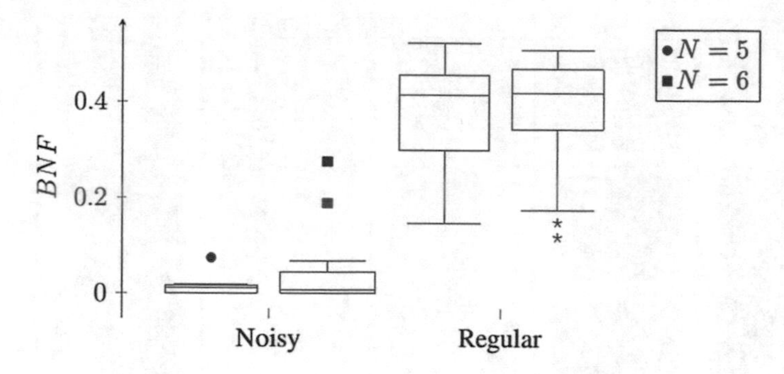

Fig. 3.12 Comparison between noisy and regular channel groups for Salinas dataset

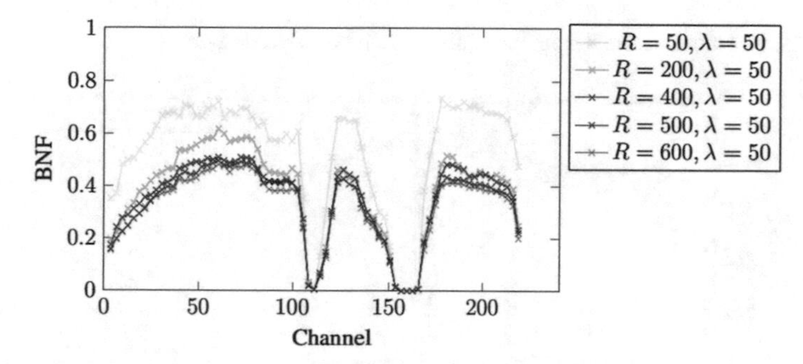

Fig. 3.13 Different #cluster parameter for Salinas dataset

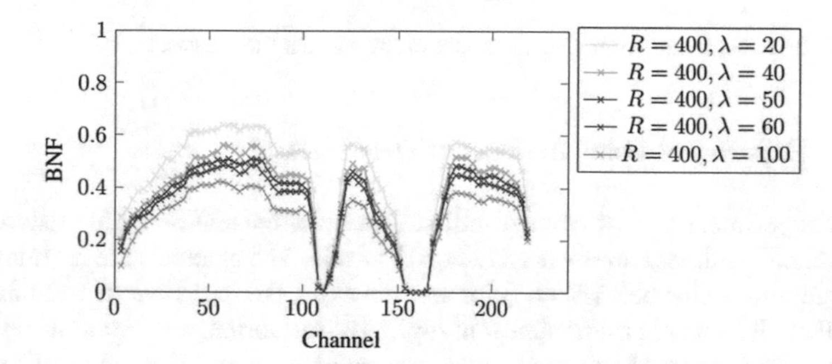

Fig. 3.14 Different λ parameter for Salinas dataset

likewise attempted another way to deal with distinguishing genuinely noisy channels. Rather than picking the following N channels for each run, we pick the following $\frac{N}{2}$ as the new grouping. The outcomes (for $N = 6$ 6, avoiding three channels each time) appeared in a similar figure were equivalent to those seen for different groups, while the BNF close to noisy transmission capacities (water-ingested transfer speeds) was more like zero.

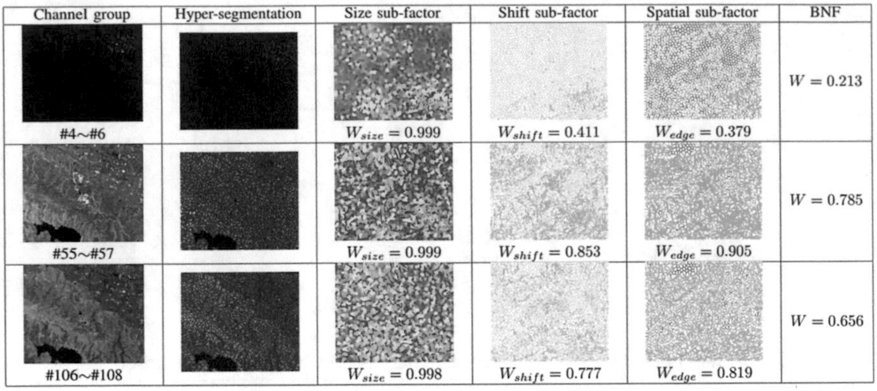

Channel group
Hyper-segmentation
Size sub-factor
Shift sub-factor
Spatial sub-factor
BNF
#4~#6
$W_{size} = 0.999$
$W_{shift} = 0.411$
$W_{edge} = 0.379$
$W = 0.213$
#55~#57
$W_{size} = 0.999$
$W_{shift} = 0.853$
$W_{edge} = 0.905$
$W = 0.785$
#106~#108
$W_{size} = 0.998$
$W_{shift} = 0.777$
$W_{edge} = 0.819$
$W = 0.656$

Fig. 3.15 Experimental results from the analysis of the Moffett dataset

$N = 3, step = 3$
BNF
Channel
0
0.2
0.4
0.6
0.8
1
0
50
100
150
200

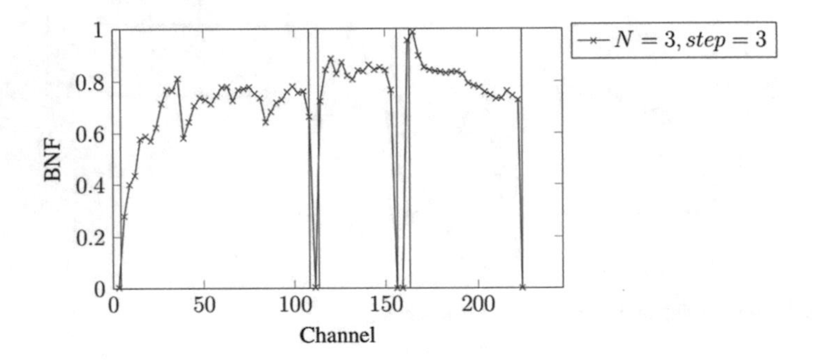

Fig. 3.16 BNF distribution over channel groups of the Moffett Field dataset

3.2.3.3 Experiment Using the Moffett Field Dataset

A third experiment was attempted utilizing the Moffett dataset. This dataset has 224 channels, with measurements 512 × 614 pixels. The channels are regrouped by three contiguous channels ($N = 3$) for assessment. Like the previous examination, the BNF of all channel groups appear in Fig. 3.16. Authoritatively pronounced noisy channels (water-engrossing wavelengths) are set apart by red lines. As can be seen, the noisy band groups for the most part had BNF <0.2, while complete noisy bands (1–3, 109–113, 153–161, and 223–224) had Band Noise Factor zero.

Figure 3.17 demonstrates the segmented outcome on some channels, while Figs. 3.16 and 3.15 demonstrate the BNF of all channels and the representation of each weight factor in some channels. As should be obvious from Fig. 3.15, the size subfactor uncovered constrained data about noise level, conceivably because of the blend of city (little articles) and landscape (huge items); while the shift and SSC subfactor both performed well to help recognize the noisy groups.

Some channels with partly noisy channels were related to a lower BNF, however none near zero. Results with various number of groups and are depicted in Figs. 3.18

Fig. 3.17 Experimental results from analysis of Moffett Field dataset with $\lambda = 4$, $R = 1,000$

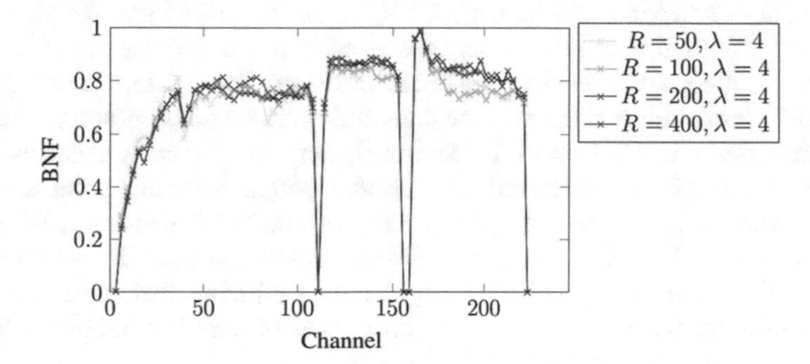

Fig. 3.18 Different #cluster parameter of Moffett Field dataset

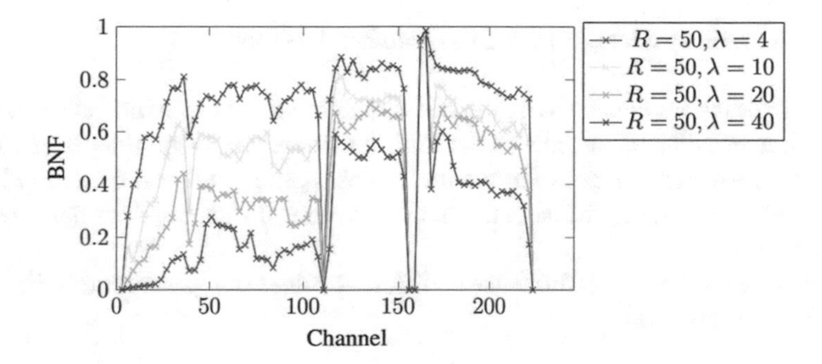

Fig. 3.19 Different λ parameter of Moffett Field dataset

and 3.19. As can be seen, the proposed structure is equipped for recognizing full noisy groups and arranging the groups as per their noise level by giving them a lower weight factor.

3.2.4 Discussion of rET Parameters

Groups of hyper segmented channel are generated, the rET algorithm, as mentioned previously on the basis of which we make computations of group BNFs. Two parameters which are predefined is used by rET algorithm, which have minor effects on the capability of our method to discern noisy channels as shown in the following experiments.[5]

3.2.4.1 Number of Clusters

The quantity of hyper-segmentation groups, or on the other hand the size of the underlying groups, primarily influenced the capacity of the rET calculation to follow landscape subtleties or little synthetic articles. Despite group size, the rET calculation will fix group limits to the genuine item limit in standard (non-noisy) channels, while the varieties in group size will consistently stay little for noisy channels. There is a slight relationship between ordinary channel BNF and starting group size, particularly when the group is huge (the quantity of clusters is deficient). We accept this is because of inadequate limit modification in the non-noisy image. The BNF differentiation among noisy and non-noisy channels, utilizing distinctive cluster size parameters, appears in Fig. 3.13. This result was received utilizing the Salinas dataset, with various segmentation appeared in Fig. 3.20.

3.2.4.2 Boundary Straightness and Gradient Factors

Curly boundaries are penalized by the rET algorithm due to specific factor that may result in longer adjustment time till the limit boundaries are straightened. Different factors are the difference between the intensities. Although it is not directly linked with the size of cluster as the adjustment of boundary straitening does not extend or move the cluster.

It is depicted in Fig. 3.14. Influence of different values of segmentation adjustment are depicted in Fig. 3.21.

3.2.5 Discussion of Weight-Subfactor Parameters

Each weight-subfactor parameter sets an edge for non-noisy channels. At the point when the proportion of the parameter surpasses one, it is cut off to one. Consequently, a λ shift too little will prompt a greater part of channels being cutoff (weighted to one) and appointing a bigger load to loud channels, since a completely noisy channel will likewise have little nonzero-group shifts. Other weight elements will all have

[5]Both experiments used the Salinas dataset with $N = 6$.

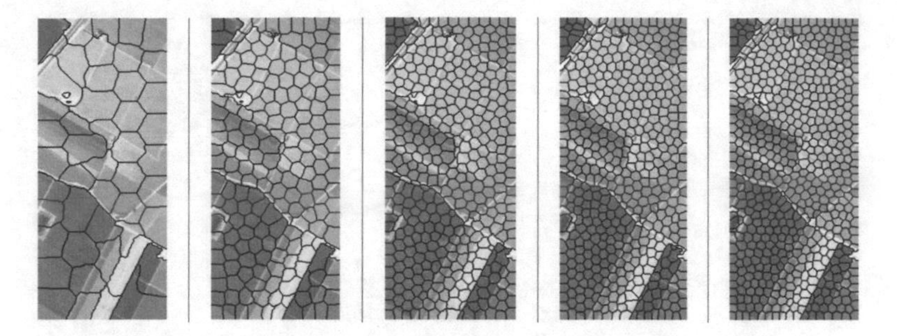

Fig. 3.20 Illustration of different segmentation-cluster size for Salinas dataset (R = 50, 200, 400, 500, 600)

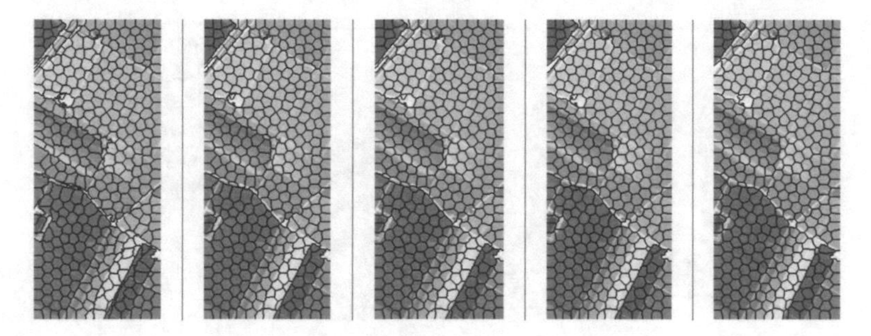

Fig. 3.21 Illustration of different edge-boundary weight factors for Salinas dataset ($\lambda = 20, 40, 50, 60, 100$)

comparable impacts. Meanwhile, a bigger λ won't remove any channel and may misidentify some typical channels as noisier than others, since there is a contrast between huge estimations of cluster variety. For instance, given that a few items are imperceptible in certain channels, non-noisy channels will have a shorter group movement with respect to channels where all objects are obvious. The excessively huge value of a λ_{shift} may make such channels have a W_{shift} penalty. A littler λ_{shift} will set the two channels to $W_{shift} = 1$. This may at present be needed in certain applications; in this manner, we propose that parameters be picked explicitly for the sort of HSI sensors utilized and dependent on other applications' specific objects.

Effect of parameter channel distribution is depicted as follows: Figs. 3.22, 3.23, 3.24 describe the influence of varying λ_{size}, λ_{shift}, and λ_{ssc}.[6]

[6]All experiments utilized the Salinas dataset with $N = 6$ and with default parameters. $\begin{cases} \lambda_{size} = 0.2 \\ \lambda_{shift} = 0.2 \\ \lambda_{ssc} = 5 \end{cases}$.

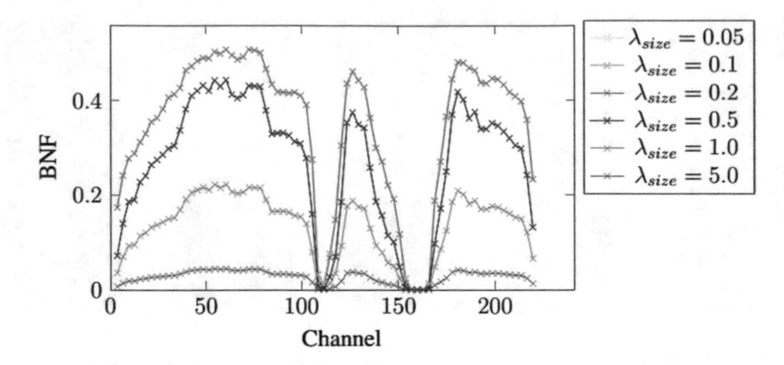

Fig. 3.22 Different λ_{size} parameters for Salinas dataset

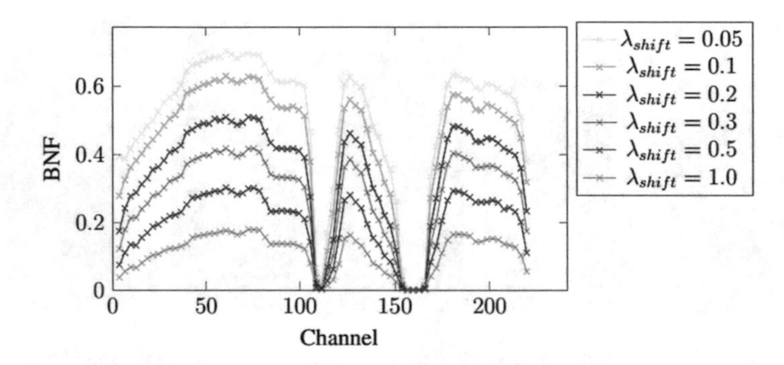

Fig. 3.23 Different λ_{shift} parameters for Salinas dataset

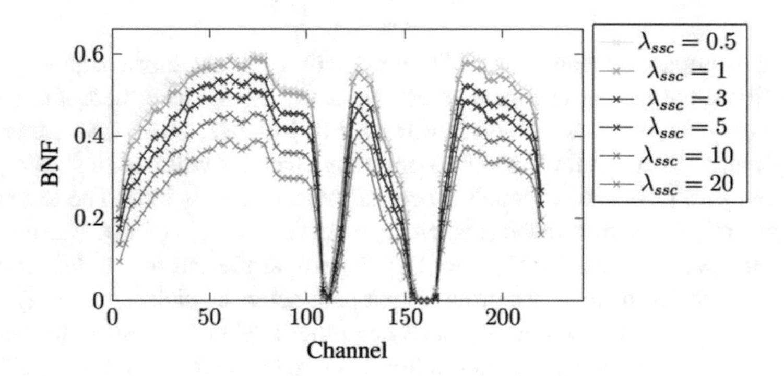

Fig. 3.24 Different λ_{ssc} parameters for Salinas dataset

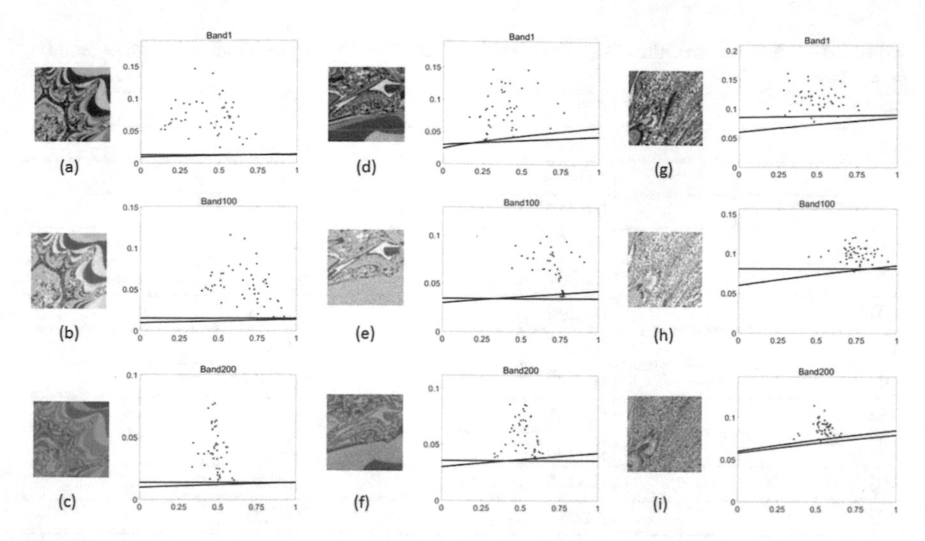

Fig. 3.25 Synthetic noisy images from fractals and the estimated noise-level function (noise standard deviation as a function of image intensity). **a–c** are from Fractal 1, with $\sigma_i = 0.01, \sigma_d = 0.01$. **d–f** are from Fractal 3, with $\sigma_i = 0.03, \sigma_d = 0.03$. **g–i** are from Fractal5, with $\sigma_i = 0.06, \sigma_d = 0.06$

3.2.6 Noise-Level Estimation

We directed noise estimation on both synthetic information and genuine HSIs. Synthetic scenes of 100×100 pixels were from a dataset by utilizing fractals to produce particular spatial patterns [17]. Noise estimation was troublesome utilizing this dataset, in light of the fact that the data contains noise like surfaces and are of low resolution. Synthetic noise was added to these images as per the noise model in [18], where both the noise factors σ_d and σ_i went from 0.01 to 0.06, with increment size of 0.01. Synthetic noise empowers the examination of the GT with the assessed noise level (ENL). We moreover did tests utilizing genuine HSIs from Salinas and Indian Pines AVIRIS information. Since we didn't have the GT of the noise level, we checked the ENLs from various hyperspectral images.

In Fig. 3.25, we demonstrate the lower envelope fitting for noise estimation with nine groups from three hyperspectral images. In (a), Fractal 3 had smooth locales, and a few examples firmly pursued the GT bend. In the meantime, the examples crossed a huge range of intensity. In this way, estimation in (a) was good. In Fig. 3.25e, most examples had exclusive requirement deviations when contrasted with the noise. The ENL didn't expand a lot, on the grounds that the examples in textureless regions held the estimation. In Fig. 3.25g, Fractal 5 had plenteous high-recurrence segments in each position, thus bringing about an overestimate.

In Table 3.1, we present the estimation of 20 groups in Factual 3. A few groups were overestimated fundamentally because of having inexhaustible surfaces and synthetic noise energies being moderately a lot of lower than those in the real image.

Table 3.1 Comparison of the GT of the noise level and the ENL, from Fractal 2, with $\sigma_i = 0.02$, $\sigma_d = 0.02$

Band	GT	ENL
1	0.0239	0.0311
21	0.0259	0.0272
41	0.0260	0.0255
61	0.0259	0.0247
81	0.0264	0.0245
101	0.0264	0.0235
121	0.0267	0.0238
141	0.0267	0.0240
161	0.0261	0.0291
181	0.0255	0.0241
201	0.0246	0.0225
221	0.0239	0.0229

We likewise applied noise estimation to actual HSIs by choosing two hyperspectral image datasets from Salinas and Indian Pines all from AVIRIS. Since the GT of the noise level in the actual HSIs was obscure, we confirmed the ENLs from nearby wavelengths. In Figs. 3.26 and 3.27, the ENLs in the contiguous groups were comparative. Water vapor ingestion groups are appeared in (c), where the ENLs were lower as contrasted and every single other example, while the GT may have traversed the focal point of the appropriation of every other example, given the speculation that the band was all noisy and contained no data. These outcomes demonstrated the point of confinement of lower envelope fitting: the noise-level capacity is thought to be lower than all examples, which is not dependable when the examples are from absolutely noisy sections.

3.2.7 HSI Classification

A further approval of the presentation and use of the proposed structure is finished by carrying out classification on hyperspectral images. In this segment, the Salinas and Indian Pines hyperspectral images are utilized to assess the Viability of the technique by choosing the subset of groups for classification. We explored AB^3C for BS by choosing the subset of groups with maximum BNF.

Unsupervised BS method [19], supervised max-relevance min-redundancy [20] method and semi-supervised band selection method [21] are utilized as standard examination strategies. Also, all channels with no choice (all bands) are applied for correlation as well.

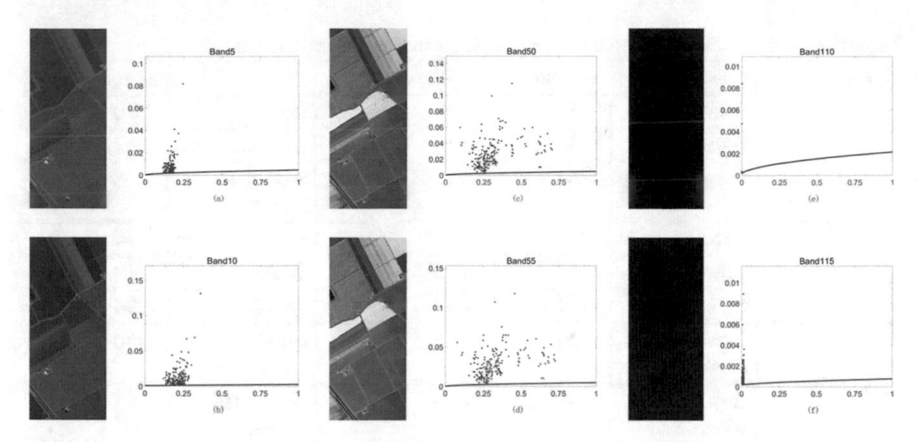

Fig. 3.26 Six bands from Salinas and the estimation result. ENLs from close bands are similar

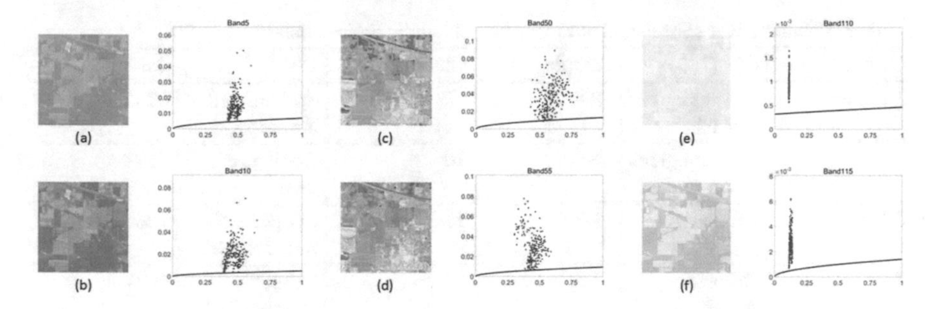

Fig. 3.27 Six bands from Indian Pine and the ENL function

To assess the viability of the chosen channels, hyperspectral images characterization utilizing loopy belief propagation and Active learning [22] is applied as a classifier.

In each of the examinations, 10% samples of each classification are utilized as training set. The well-known index overall accuracy (OA) is embraced to evaluate the grouping results quantitatively.

3.2.7.1 Classification Result Analysis

Indian Pine and Salinas datasets comprises 16 classes which are examined in Tables 3.2 and 3.3. In Figs. 3.28 and 3.29, the normal OA after effects of the five calculations on the Salinas and Indian Pine images are recorded as the quantity of chosen groups increments from 10 to 190 with interim 20. OA curve for all groups stay stable since it utilizes every one of the groups in the predefined dataset. As can be seen from the graph, the OA precision rise strongly, at that point arrive at its maximum value, lastly drop consistently. This perception shows that a specific arrangement of chosen

Table 3.2 Land-cover classes and number of pixels in the Indian Pine dataset

Class no	Class	Sample	Class no	Class	Sample
1	Alfalfa	54	9	Oats	20
2	Corn-notill	1434	10	Soybean-notill	968
3	Corn-mintill	834	11	Soybean-mintill	2468
4	Corn	234	12	Soybean-clean	614
5	Grass-pasture	497	13	Wheat	212
6	Grass-trees	747	14	Woods	1294
7	Grass-pasture-mowed	26	15	Building-grass-trees	380
8	Hay-windrowed	489	16	Stone-steel-towers	95

Table 3.3 Land-cover classes and number of pixels in the Salinas dataset

Class no	Class	Sample	Class no	Class	Sample
1	Broccoli green weeds1	2009	9	Soil vinyard develop	6203
2	Broccoli green weeds 2	3726	10	Corn senesced green	3278
3	Fallow	1976	11	Lettuce romaine 4wk	1068
4	Fallow rough pillow	1394	12	Lettuce romaine 5wk	1927
5	Fallow smooth	2318	13	Lettuce romaine 6wk	916
6	Stubble	3959	14	Lettuce romaine 7wk	1070
7	Celery	3579	15	Vinyard untrained	7268
8	Grapes untrained	11271	16	Vinyard vertical	1807

channels can give correlative data for classification. This result shows a huge number of groups may contain a lot of repetitive data and scarcely give particular valuable data for classification. Comparatively, AB^3C gets better arrangement results when the quantity of chosen groups is in the scope of [30, 190].

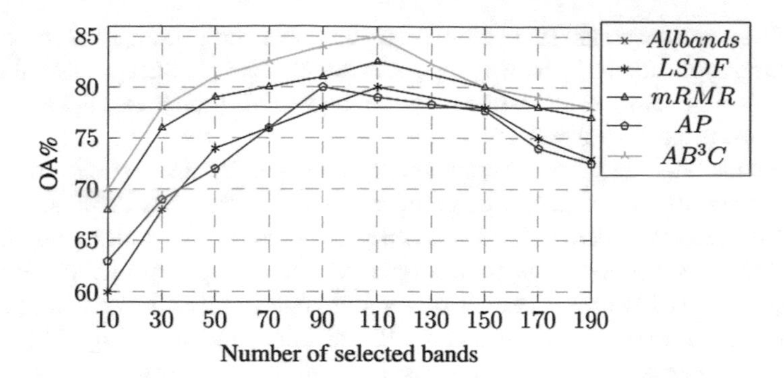

Fig. 3.28 Mean accuracies of methods on Indian Pine dataset

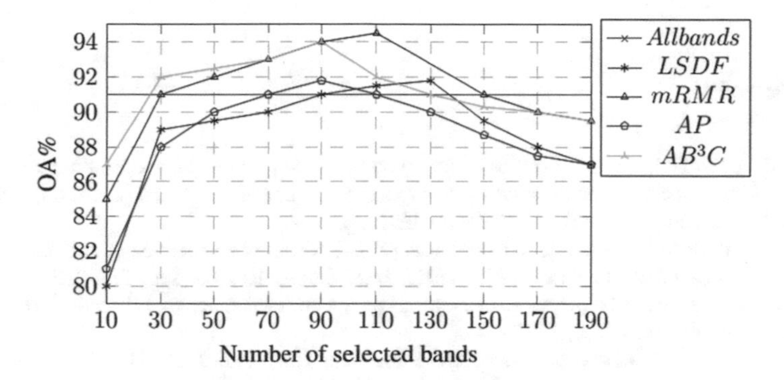

Fig. 3.29 Mean accuracies of methods on Salinas dataset

3.3 Summary of the Proposed Unsupervised Hyperspectral Image Noise Reduction and Band Categorization Method

In this examination, we presented another AB^3C structure for hyperspectral images BC that can effectively arrange HSI groups dependent on their data/quality level, consequently making the groups accessible for resulting applications/investigations, which can be considered as an essential preprocessing step. In light of the proposed strategy and the new noise model, noise level is assessed on each channel, further affirming the viability of the proposed system. To achieve high precision levels for textural and homogeneous zones of data, our technique utilizes both spatial and spectral relationships. Estimation of adaptive boundary change/development-based systems considers intraband and interband relationships. In reasonable applications, hyperspectral images gained by a progression of sensors are unavoidably diluted by noise, which may prompt huge degrees of repetitive groups sprinkled with important data. Each HSI band contains various degrees of data; in this way, the BC procedure is adaptively tuned to the data quality in each band so as to quantify data levels. BC is a significant preprocessing step before breaking down and handling HS information in resulting applications. Here, we proposed a novel, versatile BC approach for HSIs, where noise conveyance fluctuation between the various groups and the spatial data contrast inside a band are both considered.

In this methodology, band inclination is delineated by the capacity to separate each band and its data content. The group-wise A^3BC system is novel and computationally effective, which makes it valuable in practically all resulting applications related to hyperspectral images, for example, HSI denoising, capacity, transmission, and display. It is likewise a powerful preprocessing process for ensuing practical applications, for example, HSI grouping, BS, division, and unmixing. AB3C-based HSI characterization and noise estimation are additionally performed so as to show its viability in consequent applications.

References

1. Goodenough DG, Han T (2009) Reducing noise in hyperspectal data—a nonlinear data series analysis approach. In: First workshop on hyperspectral image and signal processing: evolution in remote sensing, 2009. WHISPERS'09. IEEE, pp 1–4
2. Ghamisi P, Dalla Mura M, Benediktsson JA (2015) A survey on spectral–spatial classification techniques based on attribute profiles. IEEE Trans Geosci Remote Sens 53(5):2335–2353
3. Bioucas-Dias JM, Nascimento JM (2008) Hyperspectral subspace identification. IEEE Trans Geosci Remote Sens 46(8):2435–2445
4. Veganzones MA, Tochon G, Dalla-Mura M, Plaza AJ, Chanussot J (2014) Hyperspectral image segmentation using a new spectral unmixing-based binary partition tree representation. IEEE Trans Image Process 23(8):3574–3589
5. Plaza A, Du Q, Bioucas-Dias JM, Jia X, Kruse FA (2011) Foreword to the special issue on spectral unmixing of remotely sensed data. IEEE Trans Geosci Remote Sens 49(11):4103–4110
6. Jia X, Kuo B-C, Crawford MM (2013) Feature mining for hyperspectral image classification. Proc IEEE 101(3):676–697
7. Acito N, Diani M, Corsini G (2011) Signal-dependent noise modeling and model parameter estimation in hyperspectral images. IEEE Trans Geosci Remote Sens 49(8):2957–2971
8. Sun W, Zhang L, Du B, Li W, Lai YM (2015) Band selection using improved sparse subspace clustering for hyperspectral imagery classification. IEEE J Sel Top Appl Earth Obs Remote Sens 8(6):2784–2797
9. Wang J, Wang X (2012) VCells: simple and efficient superpixels using edge-weighted centroidal voronoi tessellations. IEEE Trans Pattern Anal Mach Intell 34(6):1241–1247
10. Wang J, Ju L, Wang X (2009) An edge-weighted centroidal voronoi tessellation model for image segmentation. IEEE Trans Image Process 18(8):1844–1858
11. Levinshtein A, Stere A, Kutulakos KN, Fleet DJ, Dickinson SJ, Siddiqi K (2009) TurboPixels: fast superpixels using geometric flows. IEEE Trans Pattern Anal Mach Intell 31(12):2290–2297
12. Kerekes JP, Baum JE (2003) Hyperspectral imaging system modeling. Linc Lab J 14(1):117–130
13. Website, howpublished = www.microimages.com/documentation/tutorials/hyprspec.pdf, note = Accessed 30 Sep 2010
14. Tsin Y, Ramesh V, Kanade T (2001) Statistical calibration of CCD imaging process. In: Eighth IEEE international conference on computer vision, 2001. ICCV 2001. Proceedings, vol 1. IEEE, pp 480–487
15. Hyperspectral Computing Laboratory University of Extremadura, Cáceres, Spain, http://www.hypercomp.es/hypermix

16. Karami A, Heylen R, Scheunders P (2015) Band-specific shearlet-based hyperspectral image noise reduction. IEEE Trans Geosci Remote Sens 53(9):5054–5066
17. Plaza J, Hendrix EM, García I, Martín G, Plaza A (2012) On endmember identification in hyperspectral images without pure pixels: a comparison of algorithms. J Math Imaging Vis 42(2):163–175
18. Liu C, Szeliski R, Kang SB, Zitnick CL, Freeman WT (2008) Automatic estimation and removal of noise from a single image. IEEE Trans Pattern Anal Mach Intell 30(2):299–314
19. Frey BJ, Dueck D (2007) Clustering by passing messages between data points. Science 315(5814):972–976
20. Peng H, Long F, Ding C (2005) Feature selection based on mutual information criteria of max-dependency, max-relevance, and min-redundancy. IEEE Trans Pattern Anal Mach Intell 27(8):1226–1238
21. Zhao J, Lu K, He X (2008) Locality sensitive semi-supervised feature selection. Neurocomputing 71(10):1842–1849
22. Li J, Bioucas-Dias JM, Plaza A (2013) Spectral–spatial classification of hyperspectral data using loopy belief propagation and active learning. IEEE Trans Geosci Remote Sens 51(2):844–856

Chapter 4
Hyperspectral Image Spatial Feature Extraction via Segmentation

In the second phase, the spatial information is extracted from the Hyperspectral Image (HSI), as it is the second step toward the effective classification task as shown in Fig. 4.1. A complete description of all the HSI classification phases is depicted in Chap. 1, Fig. 1.3. This phase aims at the development of a novel unsupervised segmentation approach. Experimental results and comparison with the state-of-the-art existing segmentation approach is also presented in detail.

The major imperative goal in hyperspectral scene processing is to effectually and successfully segment the hyperspectral scene while keeping the significant vital details of the data intact. The hyperspectral image segmentation procedure is considered as a major part in the pipeline of remote sensing data understanding, investigation, and successive applications [1]. The vast expansion in spectral channels and rich spatial information not only results in deep understanding and reliability, but it also results in making the segmentation algorithm development more hard as most of the traditional segmentation algorithms either do not work or their computational complexity greatly increases due to the significant increase in spectral/spatial resolution. Furthermore, existing supervised segmentation techniques require plenty of training samples which is a major bottleneck in HSI data exploration. Unfortunately, the preparation/presence of training samples is a typical problem in pattern classification, but it becomes a major limitation when it comes to HSI specifically as it's really hard to label the HSI pixels which makes the training process really challenging. Moreover, an increase in the number of spectral channels in HSI also results in many other major issues and complications. The Hughes phenomenon (imbalance between available training data and number of dimensional) is one of them.

In the last 10 years, this vital area of HSI segmentation has been substantially explored and experimented specially in the field of supervised HSI segmentation and analysis [2–5]. In these segmentation approaches where considerable training samples are needed, various machine learning-based approaches are integrated with

L. Tao and A. Mughees, *Deep Learning for Hyperspectral Image Analysis and Classification*, Engineering Applications of Computational Methods 5,
https://doi.org/10.1007/978-981-33-4420-4_4

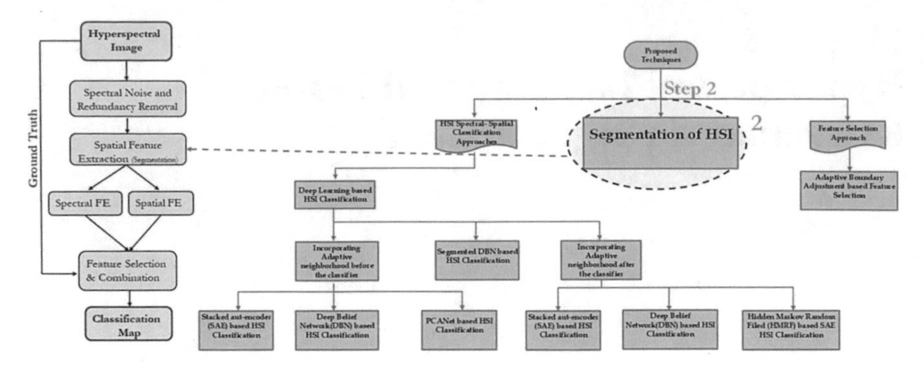

Fig. 4.1 Second developed phase aligned with the framework toward the classification of hyperspectral remote sensing images

selected/extracted features [6] from conventional algorithms. At the end, a machine learning-based classifier such as support vector machine involves training samples for classification. Various techniques established on clustering, particularly k-means and related techniques, rely on the amount of clusters given initially and don't really care about the spatial features of the data that play a vital role. In recent years, there has been increased interest in unsupervised hyperspectral segmentation algorithms. The modified watershed segmentation technique for a hyperspectral image is presented in [7]. Though the watershed approach normally results in over-segmentation, some researchers explored hierarchical segmentation [8–10]. Researchers in [8, 9] exploited minimum spanning forest (MSF). Authors in [10] explored the integration of object region detection with object region clustering resulting in a hierarchical set of image segmentation (HSEG). Some researchers approached hyperspectral image segmentation by integrating the characteristics of binary partition trees (BPT) and the theoretical background of spectral unmixing. However, the abovementioned segmentation techniques have numerous shortcomings. A major drawback is the requirement of labeled training samples in supervised segmentation, which is a vital limitation in HSI analysis. Moreover, segmentation approaches established on dimensional reduction do not completely explore the potential characteristics/features that exist in HSI.

In order to resolve the abovementioned problems, we have developed an innovative unsupervised (adapt itself according to the spectral and spatial information present) HSl segmentation algorithm, which is sensitive to the deep inter-band association between neighboring bands and spatial features without requiring any prior training samples or labeled data. The theoretical background of the developed algorithm is established on a physics-based model, and the seed initialization procedure is engrained in centroidal Voronoi tessellations. By exploiting confined structural consistency and self-similarity, the developed technique, established on the object edge movement (change of position) measures/principals, results in the efficient

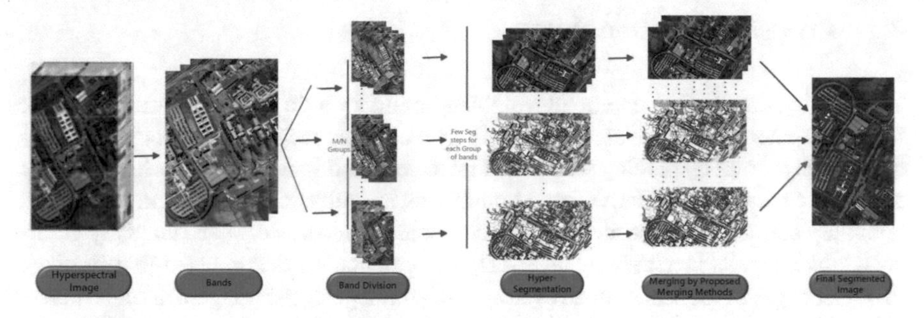

Fig. 4.2 Framework of the boundary adjustment-based weighted segmentation

segmentation of hyperspectral images. The developed algorithm resolves vital problems/limitations in the field of HSI segmentation, and it is operative not only in segmenting the HSI but also in computational complexity. The major benefits of the developed technique are summarized as follows:

1. The developed algorithm completely and effectively explores both spectral and spatial characteristics and spectral relation in neighboring channels. Furthermore, it keeps the dimensional of the HSI data intact that results in exploiting each and every channel for useful correlation and spatial dependencies.
2. The Automatic Segmentation process is established on the HSI object edge detection/movement characteristics. Each spectral band in HSI contains a unique signature for each earth surface material; we divide the hyperspectral image into B grouped channels and apply the assessment procedure on each group of channels autonomously. Weighted criteria (which is developed on many edge movement factors) is developed and used to extract valuable information enclosed in each channel for segmentation purpose.
3. Four diverse band grouping criteria are also developed that give significant performance.

The developed algorithm is applied to existing standard HSI data. Experimental results specify and show encouraging segmentation results. Moreover, it has given the new theoretical idea for segmentation. The rest of the chapter contains a detailed methodology of the developed technique. Section 4.1 explains the theory behind the developed algorithm and expresses each part of the technique in detail. Section 4.2 demonstrates the detailed experiments performed to validate the performance and efficiency of the developed framework. Finally, concluding remarks are drawn at the end.

4.1 Proposed Methodology

This section details the developed weighted boundary adjustment-based framework for HSI segmentation. The algorithm initially partitions scene into a hexagonal structure of dense areas by placing seeds to adapt to confined scene structures based on the concept of balancing of forces onto boundaries in a physical framework. Weighted boundary criteria make the algorithm to compensate and detect sharp intensity gradients, which results in achieving an effective segmentation that gives high weightage to object edges in the scene. In an even area of an image or the image with most noise, the resulting segmentation resembles Voronoi cells for each placed seeds. We have demonstrated that the developed technique is not only applicable to hyperspectral images with comparable accuracy to existing algorithms but also significantly lowers the computational complexity. Our approach is based on the following principals:

- Weighted segmentation should partition HSI into Compact regions, i.e., areas comprise unvarying shape and size, which lowers the possibility of area under-segmentation, given that the smallest segmented area is compatible with the shape of the least area that is targeted. With hexagon-based centroidal Voronoi Delaunay

Algorithm 1: Adaptive Boundary Adjustment Algorithm

```
Input : Hyperspectral image I, having pixels p with intensity vectors
        x_p, and an initial segmentation {P_k}_(k=1..K), C_{k,0} = P_k.
Output: Segmented Image I
1  bBoundaryMoved ← true;
2  i ←0;
3  while bBoundaryMoved do
4  |   i ← (i + 1);
5  |   bBoundaryMoved ← false;
6  |   for each (p, C_{k,i}) (boundary pixel p, adjacent partition C_{k,i} pair) do
7  |   |   Calculate majority class defined as (1)-(3);
8  |   |   Calculates p's local gradient grad(p) defined in (5);
9  |   |   Calculate the straightness factor defined in(6);
10 |   |   Using the above values, calculate combined attractive force
   |   |     A(p, C_{k,i}) defined in (8);
11 |   |   for All Pixels p do
12 |   |   |   compare A(p, C_{k,i}) for each neighbouring partition;
13 |   |   |   Re-assign it to the partition with largest A;
14 |   |   |   Find k such that A(p, C_{k,i}) is maximum;
15 |   |   |   put p into C_{k,i+1};
16 |   |   |   if C_reassigned ≠ C_current then
17 |   |   |   |   bBoundaryMoved ← true;
18 |   |   |   end
19 |   |   end
20 |   end
21 end
```

triangulation (CFCVDT) initialization, only the regions close to actual boundaries distort quickly. On the other hand, remaining regions remain unchanged during each repetition. In the homogeneous area of the image, most resulting segmentation sustains a roughly hexagonal shape.
- Each resulting segmented region should characterize a merely associated set of pixels. The developed dilation-based algorithm integrated with its physical forces-based boundary adjustment confirms that this limitation is constantly fulfilled.
- In the case of the nonappearance of local object boundary information, the segmented region should persist to be compact. The developed method initiates from hexagonal initialized seeds and expects no previous bias on the position of object edges. In order to exploit compactness, we have included an edge energy part that generates continuous variation in the path of the outer normal in areas of even intensity.
- When edge forces-based adjustment stops, segmented edges should match with the actual object boundaries in the scene. This involves an algorithm design with three characteristics: (1) algorithm should drop speed of its edge development in the area of image object boundaries, (2) it should be sensitive to image boundaries, and (3) it should develop plane edges. Edge decision-based proposed formulations bring a simple method to include image-based sensitivity on edge development, and contain a part for attraction shape regularization. So problems that arise in the boundary growth procedure such as object edge overpass and collision do not exist in our proposed algorithm.
- The segmentation process should allocate each pixel to a unique segmented region. Hence edge movement process should stop when two different dilating seeds are about to collide. This is automatically ensured in our algorithm.

The aforementioned considerations lead to an edge-based region growing algorithm based on the balancing of forces at every channel that we call physical model-based hyperspectral image segmentation, which aims to sustain and evolve the edges between the allocated regions until an object boundary is encountered. The proposed approach can be conceptually divided into the following main steps:

1. Divide initial hyperspectral M channel image into $\frac{M}{N}$ N-channel groups.
2. For each Group of channels, perform the following steps:
    a. Place m initial seeds for m classes in the image.
    b. Based on the seeds, divide the region into m initial hexagonal segments/clusters.
    c. Repeat over the subsequent elementary stages till no additional development is conceivable, i.e., no pixel going to a diverse segment.
        - Compute the class centroid of each initial divided segment/cluster.
        - Compute the gradient of current boundary pixels of each segment/cluster.
        - Compute the straightness factor for each cluster boundary.
        - Based on the equation, compute the measurement of similarity, i.e., the force of attraction/repulsion between each class centroid and its edge parts.

- Make the edges to progress built on the similarity measure.

3. For each band cluster, allocate weight to every edge pixel conferring to convinced criteria, to assign pixels which reflects an actual edge.
4. Merge N channel to obtain a weighted segmentation of HSI.

We will discuss each process in detail. The overall structure of the developed algorithm is depicted in Fig. 4.2 and is conceptually presented in Algorithms 1 and 2. Each part of the framework is detailed as follows.

4.1.1 Preprocessing Toward Initial Segmentation

To detect object borders/edges, the original hyperspectral image data consisting of M bands is grouped into $\frac{M}{N}$ G-band groups, then Algorithm 1 is evaluated on each resulting group autonomously and then results of each group are merged conferring to Algorithm 2. The process of grouping the adjustment channels contributes a vital part. In the absence of the grouping, noise that exists in each band may dominate the signal which results in ignoring the actual border/edge information in each band. Therefore, grouping plays an important role in suppressing the pure noise present in each channel.

The HSI data comprises pixels $p_{i,j}$ (with $i \in [1, width]$, $j \in [1, height]$) and is partitioned into a group of similar pixels ("cells") $C_i \in \mathbf{C}$ where each pixel in a particular group is $p \in C_i$; the properties/features of an individual pixel are characterized as a real-valued high-dimensional normalized vector $x_p \in [0, 1]$. Provided a band group, we characterize individual pixels as N-dimensional vectors, and apply the abovementioned Algorithm 1 to discover all the segmentation regions in the given hyperspectral image. More precisely, the cluster set $\mathbf{C} = (C_1, \ldots, C_R)$ of hexagonal segments will be obtained which are approximately of the same size and shape and a boundary definition $\mathbf{E} = \{p|N(p) > 1\}$. For step 1, we have selected Lloyd's Hexagon algorithm [11] and CFCVDT [12].

4.1.1.1 Initial Seed Placement

The objective of the first stage of the proposed algorithm is to place k initial seeds based on the assumption that there are k classes in the given HSI and generate small hexagon-shaped segmentation regions which are approximately of even contour and size. The number of classes in an image mainly rests on the complexity of the image. Our choice for stage one is based on Lloyd's Hexagon algorithm [9] and CFCVDT [10], an active and efficient approach for partitioning and clustering image pixels. The CFCVDT-based partitioning approach is capable of generating approximately 10,000 tiny hexagonal image segments based on the seeds in less than a second, and its run time is not dependent on the size of the image. So CFCVDT can be effectively applied to Hyperspectral Images. Overall, CFCVDT results in small hexagonal-shaped segments, independent of the image spatial or spectral information.

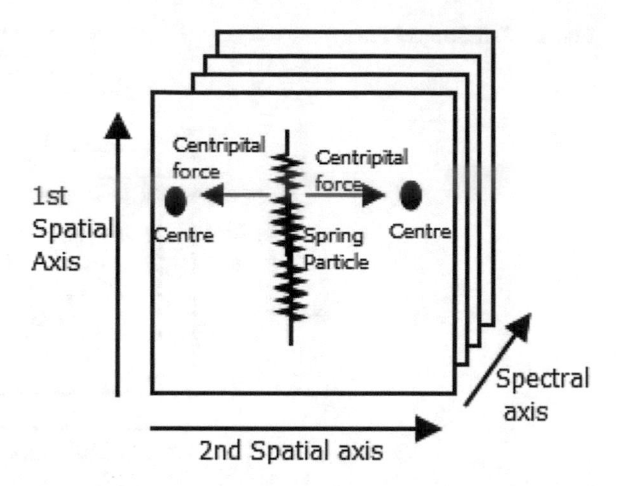

Fig. 4.3 Representation of a boundary adjustment taking into account both spectral and spatial information of the image

Specifically, the initial phase is an initialization procedure as calculated segments are utilized as an early segment for the next stage. By choosing a hexagonal shape, the partition boundary is more flexible as compared to partitioning with other shapes, which gives a strong horizontal/vertical tendency over the final boundary.

4.1.2 Boundary Model

An intuitive explanation of the boundary adjustment in a physical model is presented in Fig. 4.3.

The boundary is formed by a series of particles lying in a flat surface, linked via springs which exert forces in multiple directions (due to multiple channels) to maintain the link straight. Each particle is attracted toward adjacent partitions' class centers; the magnitude of forces is related to the spatial difference between the particle's position and partitions' spatial information. The surface has a variable friction coefficient; some certain area with higher spatial gradient is set to be coarser. Ideally, the particles will rest in positions such that their linkage forms a perfect partitioning of the surface image, where each particle lies in coarser (boundary) positions, and spatial information in each partition is uniform. By simulating the particles' motion, we can determine an adequate partitioning of underlying HSI. An example equilibrium of the above physical model is presented in Fig. 4.4.

The initial concept behind quality segmentation is to formulate an iterative optimization to maintain and evolve the boundary by tri-factor, to coincide with the object boundaries in HSI.

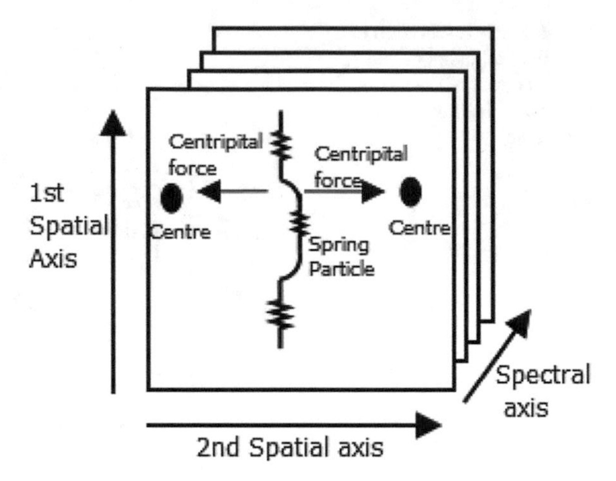

Fig. 4.4 Balancing of forces

4.1.2.1 Major Class Computation

Primarily, there are R number of clusters positioned on the HSI scene. These clusters are of hexagonal shape. As a result of placing the clusters randomly, the scene consists of random segments with the possibility of each segment containing more than one as preliminary clustering is produced independent of any spectral signature of spatial correlation. Hence logically, the subsequent phase is the calculation of the main class that exists in that particular cluster. The main class is evaluated on the spatial–spectral contextual (SSC) value in each preliminary cluster. For each partition P_k, we choose g_k to represent the majority class of its member pixels. This resultant class is designated through a repetitive removal procedure. Primarily, $C_{k,0} = P_k$. The spatial–spectral contextual value and standard deviation (SD) are computed for each hexagonal-shaped existing cluster k:

$$\begin{cases} n_i = |C_{k,i}| \\ \overline{x_{k,i}} = \frac{1}{n_i} \sum_{p \in C_{k,i}} x_p \\ \sigma_{k,i} = \sqrt{\frac{\sum_{p \in C_{k,i}} (x_p - \bar{x_{k,i}})}{n_i}} \end{cases} \tag{4.1}$$

Each minor is then removed repetitively by utilizing constant ε:

$$C_{k,i+1} = \left\{ p \in C_{k,i} \middle| \left| x_p - \overline{x_{k,i}} \right| > \varepsilon \times \sigma_{k,i} \right\} \tag{4.2}$$

The mainstream class set is declared through repetitive constant I: $M_k = C_{k,I}$, and the major class is based on the mean of these pixels:

$$g_k = \frac{1}{|M_k|} \sum_{p \in M_k} x_p \tag{4.3}$$

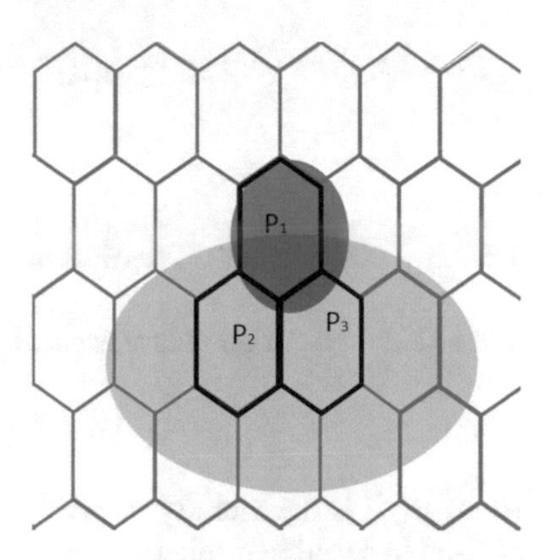

Fig. 4.5 Demonstration

This complete repetitive removal procedure can guide in identifying the original bounded object in the unadjusted segment. Consider the following example.

The majority color g_1 for partition P_1 is purple, regardless of calculation; however, partitions P_2 and P_3 covered more than one object, and the naive average color is a mix of two colors. However, the minority-colored pixels is gradually expelled from $C_{k,i}$ and finally M_2, M_3 contain only pixels from the majority object, and majority color g_2, g_3 is pure green (Fig. 4.5).

Note that a boundary pixel should move away from a uniformly colored partition, especially when the color g_k is close to x'_p (since this particular pixel is highly likely to be a part of the partition's underlying object); therefore, we let the color difference between the boundary pixel and majority color $|x'_p - g_k|$ contribute to the partition's attraction to the boundary.

4.1.2.2 Gradient Calculation

As in a natural object segment, intensity characteristics tend to be similar, which results in a significantly large gradient value at the boundaries of the actual object segment. This phenomenon can be observed during the experimentation phase. This applies that a stronger gradient value means that there is a high possibility of the presence of a potential boundary. The pixel $p = p(i, j)$ can now be visualized in the Cartesian coordinate system distinct on the scene pixels. Therefore, the specific intensity now corresponds to the vector element in the xy coordinate system, $x(i, j) = x_{p(i,j)}$. With this data in place, we can present a gradient value for each pixel value $p(i, j)$.

$$|\nabla x(i, j)| = \left|\left(\frac{\partial x(i, j)}{\partial i}, \frac{\partial x(i, j)}{\partial j}\right)\right| \tag{4.4}$$

Since pixels are demarcated individually, additional definition of the slope of the pixel location utilizing the 4-path kernel technique is

$$grad(p_{i,j}) = \frac{1}{4}\{\|x_{p_{i,j}} - x_{p_{i+1,j}}\| + \|x_{p_{i,j}} - x_{p_{i,j+1}}\| + \|x_{p_{i,j}} - x_{p_{i-1,j}}\| + \|x_{p_{i,j}} - x_{p_{i,j-1}}\|\} \quad (4.5)$$

4.1.2.3 Straightness Calculations

The straightness of boundary aspect $\tilde{n}_k(p)$ is characterized as follows, same as [13]:

$$\tilde{n}_k(p) = |B(p) \setminus (P_k)| \quad (4.6)$$

Here, $B(p)$ is the vicinity of pixel p. Generally, the constant value of radius R is utilized to describe a spherical neighborhood:

$$B(p(i, j)) = \{p(i', j')|\sqrt{(i - i')^2 + (j - j')^2} \leq R\} \quad (4.7)$$

One must observe here that two neighboring hexagonal partitions contribute to the same amount of neighboring parts; this applies to the fact that the straightness factor is equal for both of these sectors $\tilde{n}_k(p)$. If these two factors are distinct, then this factor will have more weightage for the sector with less neighboring parts, in other words, the hexagonal segment is enclosed by the curved edge (that is P_2, and so, $\tilde{n}_2(p) > \tilde{n}_1(p)$). The edge tries to be more straight if this factor have some weightage in the attraction force later .

4.1.2.4 The Combined Attraction Force

So the resulting integrated attraction force for each addressed pixel to its neighboring sector is described as follows:

$$A(p, P_K) = \sqrt{|x_p - g_k|^2 + \lambda\tilde{n}_k(p)|Grad(p)|} \quad (4.8)$$

where, as formerly described,

- x_p is defined as the intensity at the edge pixel,
- g_k contains the major intensity value,
- $\tilde{n}_k(p)$ is the factor for the weightage of edge to be straight,
- $|Grad(p)|$ is the value of the intensity gradient at specified pixel p

and λ is a self-described weight for smoothing the edge.

4.1.2.5 Adjustment and Termination

If provided with the specified sector $\{P_k\}$, the edge pixel can be described as all those pixels which have neighboring pixels that fit into a distinct sector. For edge pixels, we describe "neighboring sectors" to be sectors with neighboring pixels.

$$Adj\_pixels(p(i,j)) = \{p(i \pm 1, j \pm 1)\} \quad (4.9)$$

$$Adj\_partitions(p) = \bigcup_{p_a \in Adj\_pixels(p)} \{P(p_a)\} \quad (4.10)$$

Throughout each repetition, the edge pixels are assigned to different sectors based on the attraction force, which is evaluated as follows:

$$P'(p) = arg \max_{P_k \in Adj\_partitions(p)} A(p, P_k) \quad (4.11)$$

and then the resulting edge is re-evaluated based on the sector after each repetition.

The repetitive modification procedure dismisses if the edge converges. After each repetition, if no edge pixel is attuned (all segmentation is the same, i.e., $P'_k = P_k, \forall k$), then we infer that these are the actual boundaries.

4.1.3 Channel–Group Merge Criteria

The final evaluated boundaries are highly based on the segmentation factors, i.e., size of a cluster, cluster shift, and spectra–spatial contextual factors contribute toward the final segmentation. For every band group, the product of all the resulting factors is taken to formulate the concluding pixel value $W_m(i,j)$. We have $W = W_{size} \times W_{shift} \times W_{ssc}$ where W_{size} shows variance of cluster dimensional after adjustment, W_{shift} is the mean distance the clusters have moved during the modification procedure, and W_{ssc} is the SSC change at the clusters' edges. The concluding weighted boundary $wb(i,j)$ is an additional integration of all weights from distinct band groups .

There are several possible ways to achieve this merge:

1. Cut-off Binary: this pixel is a boundary pixel as long as one of the channel groups reports weight larger than Threshold T, or

$$wb(i,j) = \begin{cases} 1, & if\ \exists m, W_m(i,j) > T \\ 0, & otherwise \end{cases} \quad (4.12)$$

Algorithm 2: Hyper-Segmentation

Input : Hyperspectral image $\mathcal{I}$, with M channels
Output: Final Segmentation Result after Merging
1 Divide M channels into $\frac{M}{N}$ G-channel groups;
2 **for** $i \leftarrow 1$ *to* G **do**
3   Apply *Algorithm*1 on each channel Group;
4   save the result for each Channel Group;
5 **end**
6 Apply Merge criteria's defined in (12)-(14) on resulted G images;

---

2. Cut-off Majority Voting: this pixel is a boundary pixel as long as the majority of the channel groups report weights larger than Threshold T, or

$$wb(i,j) = \begin{cases} 1, & if \ |\{m|W_m(i,j) > T\}| > \frac{N}{2M} \\ 0, & otherwise \end{cases} \quad (4.13)$$

3. Cut-off Sum: this pixel has a summation of channel group weights, normalized to [0, 1], for later cut-off.

$$wb(i,j) = \sum_{m=0}^{N/M} W_m(i,j) \quad (4.14)$$

4. Threshold Sum: this pixel has a summation of all channel group weights that are larger than Threshold T, and then it is normalized to [0, 1].

$$wb(i,j) = \sum_{m=0}^{N/M} \begin{cases} W_m(i,j), & if \ W_m(i,j) > T \\ 0, & otherwise \end{cases} \quad (4.15)$$

Algorithm 2 summarizes the implementation of the whole segmentation process.

4.2 Experimental Approach and Analysis

This part of the chapter describes the experimental setup and results of the developed segmentation approach. The quality of the segmentation is also compared with the state-of-the-art existing technique named as spectral unmixing-based binary partition tree (BPT) [14].

4.2.1 Hyperspectral Datasets

Two well-known and regular HSI datasets Pavia University and Indian Pine are selected in demonstrating the performance of the algorithm. These datasets are selected because of their complexity and usage as these HSI scenes have widely been used by researchers all over the world for the past 20 years to evaluate their algorithms. These datasets have earned the position as a benchmark for HSI algorithm performance evaluation. A detailed description of each dataset is described in Chap. 2, Sect. 2.3.

4.2.2 *Adjustment of Weight Factors*

This section briefly highlights the influence of the combined weight factor in each band group by integrating the result of weight factor for each band group and then combining the result of all the band groups to form a single resulted object boundary. In this process, other criteria are inactivated. The resultant weighted segmentation result are also compared with the prevailing state-of-the-art segmentation technique to demonstrate the features of the developed approach. The columns of Table 4.1 demonstrate the outcome with one of the three weight subfactors showing results with only one of the weight subfactors empowered, and the rest are inactivated. As can be seen in Table 4.1, the cluster size factor supports and shows the maximum response at the complex tiled region of HSI Pavia University. It also plays a very important role in the Indian Pine regions with big tiles as it permits the cluster to grow flexibly.

Furthermore, the shift factor largely supports to remove the extra edges produced in the homogeneous regions. The clusters are unable to make any movements in these areas.

Lastly, the SSC variance factor also supports to correct edges shaped in homogeneous regions and to reinforce actual object edges. This effect can be seen in the segmentation region of the Pavia image, where the buildings are segmented on the lower left side.

One must observe that the developed segmentation algorithm is for general purpose but in order to segment the image that contains some specific properties, some algorithm parameters may need to be fine-tuned.

Table 4.1 Experimental results on the Pavia and Indian Pine datasets with different parameters

SSC Sub-Factor only	Shift Sub-Factor only	Size Sub-Factor	Combined Factors	Original Image

4.2.3 Grouping and Merging Methods

This section comprehensively discusses and presents the whole process employed to perform and investigate the results of segmentation on the abovementioned HSI dataset. Precisely, we present the phases followed to acquire the accurate segmentation. Initially, given HSI data is separated and connective bands are grouped to form a number of grouped bands where each group contains an equal number of bands. We experimented on a group of 5, 10, 15, and 20. Extensive experiments are performed and the algorithm is implemented on each of the band groups, and the final results for each group band are integrated through the four developed merging criteria. It was experimentally observed that our algorithm provides the best results on Pavia HSI if 103 bands of Pavia are grouped as 7 in each group resulting in a total of 15 groups. On the other hand, grouping 5 bands each in the Indian Pine dataset gives fine segmentation regions. As stated earlier, various strategies are developed for integrating the resulting boundaries from several band groups, ranging from the simplest binary cut-off to the voting-based technique.

The naïve cut-off binary merging technique is incapable of producing significant segmentation results related to other developed merging techniques. The main reason is that in some cases the information about the actual boundary is not sufficient which results in very low signatures of the boundaries in each band. At the time of merging this information of weak boundary signature from each band, a significant portion of the noise is integrated into the resulting merged boundary. The second proposed voting-based technique is also unable to perform well as it is possible that one signature is strong in any one of the frequency range of the spectrum but very low in the other range, hence it could not be included in this voting-based criteria.

On the other hand, the cut-off sum and threshold sum (TS) techniques achieved enhanced performance for the HSI scene and came up with quality segmented regions. The influence of randomly generated segments in the uniform regions can further be reduced if a cut-off sum is applied before applying TS. The outcome of the proposed merging techniques is compared in Table 4.1.

4.2.4 Experimental Results and Comparison

In the detailed experimentation phase, a distinct contribution of each segmentation criteria, i.e., cluster size, cluster shift, and spatial–spectral contextual (SSC) information factors, is analyzed, as depicted in Table 4.1. The segmented regions of HSI are presented in Table 4.1 and Fig. 4.7. One must observe that spatially important regions like buildings, university areas, parking allotted places, and other obvious buildings are accurately segmented with clear boundary indications. Small regions which have very less allotted pixels are also well detected.

The evaluation of the Pavia University hyperspectral dataset with the existing famous BPT algorithm is depicted in Fig. 4.6. While comparing it with the prevailing

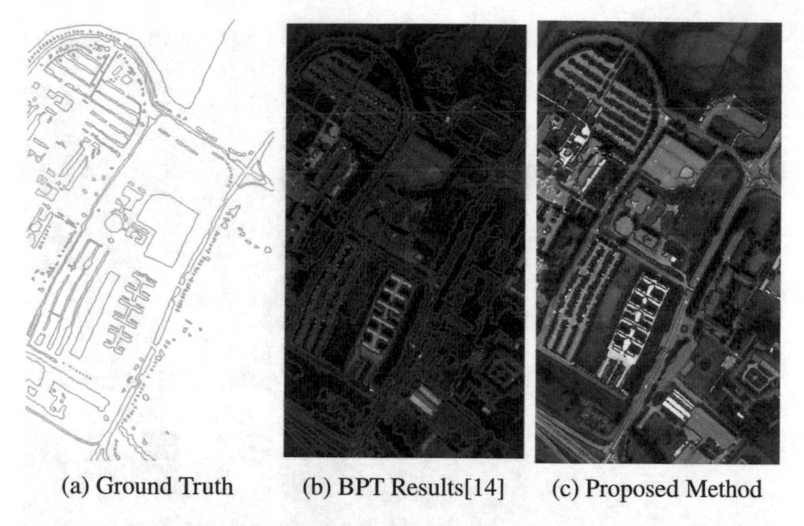

Fig. 4.6 Comparison of the entire Pavia University scene

BPT [14] method, it is obvious from Fig. 4.7 that part of the HSI scene comprising the parking region (lower middle) is entirely and precisely segmented by the developed approach. However, the BPT method is only able to segment a partial part of it and is unable to segment the small regions. Likewise, in the central leftward portion of the scene, the developed algorithm precisely gave the quality segmentation results even on the minor regions including the rounded portion. On the other hand, the BPT algorithm is unable to detect these regions well. A similar kind of performance of the developed approach can be seen in other datasets. Likewise, the proposed technique also successfully gave quality segmentation results on part of the Pavia scene which is not included in the Ground Truth. For instance, the developed method has efficaciously resulted in the quality segmentation of the middle right portion of the Pavia scene. This part is not included in the ground truth.

4.2.5 Evaluation Measures

In the cases where we need to classify the image on the basis of objects present in the scene, segmentation is the tool utilized for generating those objects for example object based image classification and segmentation. Over-segmentation is the result of dividing a single object into more than one object during the segmentation process. When the segmentation produces objects that encompass more than one validation object, under-segmentation occurs. These two, i.e., over- and under-segmentation are considered one of the major errors in the segmentation field. In order to handle these two problems in segmentation, metrics-based qualification procedures have been developed to effectively show the accuracy of the resulting segmentation. Over-

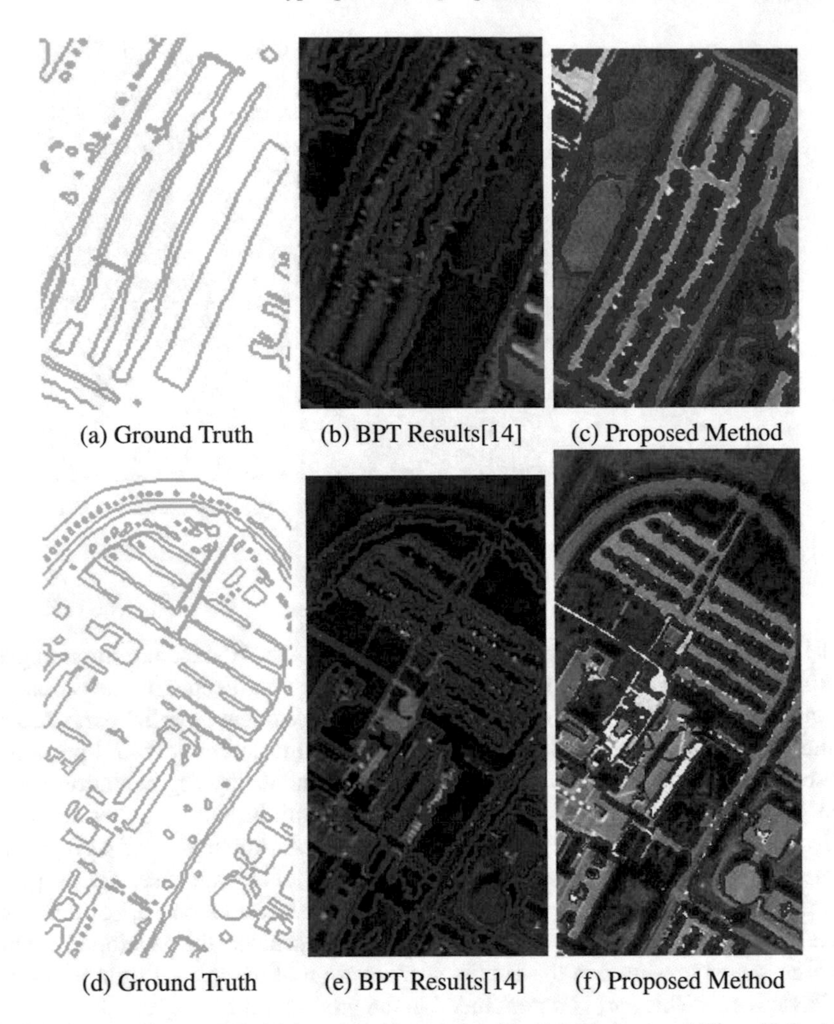

Fig. 4.7 A magnified section of the segmentation result in Fig. 4.6. The lower middle and left middle parts are magnified for comparison. The proposed method performed really well and segmented even the small areas in these regions as compared to the existing method

segmentation metrics are mainly based on the ratio of the overlapping area to the validation object area. Under-segmentation metrics, on the other hand, are mainly based on the ratio of the overlapping area to the classified object area. Intuitive ways to combine these two metrics include the sum and root mean square. The F-measure has also been adopted for this combination The following criteria were applied for quantitative performance evaluation of the segmentation results.

Probabilistic Rand Index (RI): It calculates the fraction of pairs of pixels whose labels are consistent between the ground truth and the resulting segmentation [15].

Variation of Information Metric (VOI): It describes the distance between two segmentations as the average conditional entropy of a segmentation given the other [16], in our case it's the Ground Truth. It calculates the amount of randomness in the segmentation which cannot be explained by the other. Both evaluation criteria must be measured together as they conclude each other; RI should have a high value and at the same time, VoI should have a less value.

Inputs to the evaluation measures are the selected classes of the segmented image Fig. 4.6 and the ground truth. Figure 4.7 shows the resulting measures which clearly shows the accuracy of the proposed method.

4.3 Summary of the Proposed Hyperspectral Image Spatial Feature Extraction via Segmentation Method

In this part of the book, a novel framework was designed and discussed in detail. This developed framework effectively performs segmentation on the hyperspectral image without the requirement of prior labeled training samples, i.e., unsupervised segmentation. It is formulated on a flexible edge movement-based algorithm which takes the full lead from all the available spectral channels. The developed method earns full lead from spectral and spatial formations available from the scene. The distribution of information within various channels is extricated based on tri-factor weight criteria. These criteria are rooted from the measurement of the variation of boundaries. The process of segmenting the scene segmentation process is flexibly adjusted with the information present in each channel by segregating the hyperspectral scene into multiple channel groups. The offered framework discourses the vital issue of HSI segmentation in an interconnected style, i.e., by combining the features of the weighted edge movement framework.

The developed approach has been tested on challenging AVIRIS and ROSIS HSI representing two perspectives, natural sceneries and urban scenes. Both possess unique spectral bands and spatial resolutions. Normally, the subsequent segmented areas of the scene are vital and critical for the subsequent investigations and applications. The four devised merging result criteria play a very important role in finalizing the segmentation results and give a final vital touch to the final segmentation. More specifically, first extricating the edge information using tri factor weight criteria from each distinct set of channels based on the distinct spectral and spatial information each channel group have and then integrating this information from the developed 4 merging criteria's results into a very strong boundary which greatly matches the actual object boundaries. Furthermore, it is important to mention that developed approach works greatly even on rough noisy scenes and on high intensity scenes.

References

1. Ghamisi P, Dalla Mura M, Benediktsson JA (2015) A survey on spectral–spatial classification techniques based on attribute profiles. IEEE Trans Geosci Remote Sens 53(5):2335–2353
2. Dópido I, Villa A, Plaza A, Gamba P (2012) A quantitative and comparative assessment of unmixing-based feature extraction techniques for hyperspectral image classification. IEEE J Sel Top Appl Earth Obs Remote Sens 5(2):421–435
3. Guo B, Gunn SR, Damper RI, Nelson JD (2008) Customizing kernel functions for SVM-based hyperspectral image classification. IEEE Trans Image Process 17(4):622–629
4. Plaza A, Benediktsson JA, Boardman JW, Brazile J, Bruzzone L, Camps-Valls G, Chanussot J, Fauvel M, Gamba P, Gualtieri A et al (2009) Recent advances in techniques for hyperspectral image processing. Remote Sens Environ 113:S110–S122
5. Massoudifar P, Rangarajan A, Gader P (2014) Superpixel estimation for hyperspectral imagery. In: 2014 IEEE conference on computer vision and pattern recognition workshops (CVPRW). IEEE, pp 287–292
6. Serpico SB, Moser G (2007) Extraction of spectral channels from hyperspectral images for classification purposes. IEEE Trans Geosci Remote Sens 45(2):484–495
7. Tarabalka Y, Chanussot J, Benediktsson JA (2010) Segmentation and classification of hyperspectral images using watershed transformation. Pattern Recognit 43(7):2367–2379
8. Bernard K, Tarabalka Y, Angulo J, Chanussot J, Benediktsson JA (2012) Spectral–spatial classification of hyperspectral data based on a stochastic minimum spanning forest approach. IEEE Trans Image Process 21(4):2008–2021
9. Tarabalka Y, Chanussot J, Benediktsson JA (2010) Segmentation and classification of hyperspectral images using minimum spanning forest grown from automatically selected markers. IEEE Trans Syst Man Cybern Part B (Cybern) 40(5):1267–1279
10. Tarabalka Y, Tilton JC, Benediktsson JA, Chanussot J (2012) A marker-based approach for the automated selection of a single segmentation from a hierarchical set of image segmentations. IEEE J Sel Top Appl Earth Obs Remote Sens 5(1):262–272
11. Lloyd S (1982) Least squares quantization in PCM. IEEE Trans Inf Theory 28(2):129–137
12. Vincent P, Larochelle H, Lajoie I, Bengio Y, Manzagol P-A (2010) Stacked denoising autoencoders: learning useful representations in a deep network with a local denoising criterion. J Mach Learn Res 11(Dec):3371–3408
13. Wang J, Wang X (2012) VCells: simple and efficient superpixels using edge-weighted centroidal voronoi tessellations. IEEE Trans Pattern Anal Mach Intell 34(6):1241–1247
14. Veganzones MA, Tochon G, Dalla-Mura M, Plaza AJ, Chanussot J (2014) Hyperspectral image segmentation using a new spectral unmixing-based binary partition tree representation. IEEE Trans Image Process 23(8):3574–3589
15. Du Q, Fowler JE, Zhu W (2009) On the impact of atmospheric correction on lossy compression of multispectral and hyperspectral imagery. IEEE Trans Geosci Remote Sens 47(1):130–132
16. Meilă M (2005) Comparing clusterings: an axiomatic view. In: Proceedings of the 22nd international conference on machine learning. ACM, pp 577–584

Chapter 5
Integrating Spectral-Spatial Information for Deep Learning Based HSI Classification

This chapter presents a detailed analysis and development of deep learning (DL) based techniques for hyperspectral image classification. This is the third phase in our developed framework as shown in Fig. 5.1. A complete explanation of each stage is illustrated in Chap. 1, Fig. 1.3. In this phase, three different DL-based algorithms are developed for HSI classification. The proposed DL-based methods include two key techniques—first, spatial features are extracted through a hyper-segmentation, which is developed in the second phase of the thesis where dimensional and outline of the segmented regions can be modified conferring to the actual object-level spatial regions and that also comprises of pixels that are not only spatially connected but also have similar spectral signatures, secondly, object-level classification by exploiting DL architectures based decision fusion method is developed. Three different DL architectures stacked auto-encoder, Deep Belief Network and PCANet are investigated and the performances of the algorithms are evaluated and compared against the existing state-of-the-art HSI classification techniques.

As discussed in detail in Chaps. 1 and 2, operative and precise hyperspectral image (HSI) exploration have become significantly vital through rapid developments in remote sensing technology. More advanced remote sensing technology has the capability to acquire useful earth's surface data consisting of not only hundreds of spectral bands but also equipped with a comprehensive spatial resolution of the captured scene. This wealth of amusing spectral and spatial data can result in much precise classification and analysis of the scene [1]. Furthermore, modern and more recent remote sensing sensors are equipped with the ability to obtain enormous spatial resolution [2]. It makes the identification of small spatial regions, hidden objects possible in HSI.

The above-mentioned developments make the HSI effective in many fields ranging from surveillance [3], agriculture [4], mineralogy [5], astronomy [6], and environmental sciences [7, 8]. However, the wealth of spatial and spectral information comes with practical and theoretical challenges such as: curse of dimensionality [9],

L. Tao and A. Mughees, *Deep Learning for Hyperspectral Image Analysis and Classification*, Engineering Applications of Computational Methods 5,
https://doi.org/10.1007/978-981-33-4420-4_5

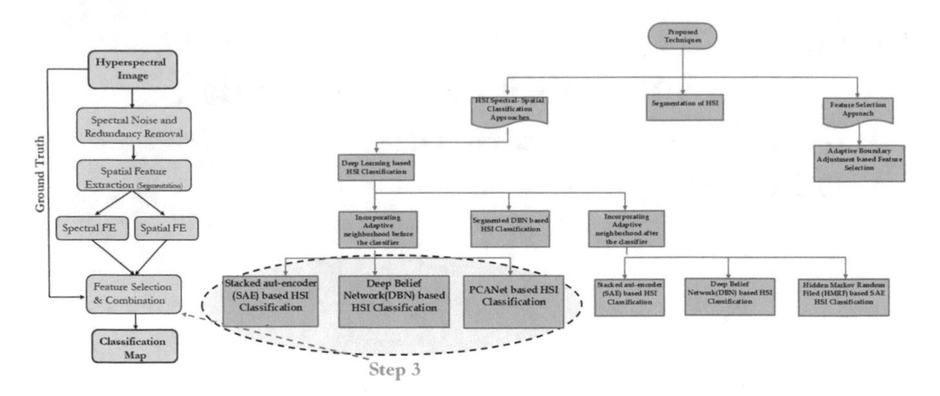

Fig. 5.1 Third phase aligned with the framework toward the classification of hyperspectral remote sensing images

Hughes phenomena [10], limited training samples, complex statistical characteristics and high computational problems.

To address the dimensionality issue, the existing HSI classification methods have primarily worked on band selection (FS) and/or band extraction (FE) to give the specified classifier as an input [11] for classification task. In these two feature selection/extraction approaches, feature selection techniques involve selecting the best subclass of channels based on some predefined criteria, while feature extraction approaches convert the numerous spectral channels into the into fewer dimensional, mostly in 3 channels. One of the renowned classifier, Support Vector Machines (SVMs) gave a significant classification results on HSI with a small set of training data [11, 12]. Primarily, techniques based on feature extraction/dimensional reduction procedures such as Principal Component Analysis (PCA) and Independent Component Analysis (ICA) end up losing the valuable data which greatly hurts the subsequent classification results. Furthermore, these FE approaches also end up losing the spatial features which involve local object boundaries.

In recent times, the hyperspectral image analysts and scientists have recognized and appreciated the significance and vital role of integrating the spatial structural features combined with spectral information for the promising analyses and classification of the HSI scenes [13, 14]. It is because newly advanced developed remote sensing sensors can not only provide rich spectral data but can also capture prominent spatial structures. Spatial information is as important as the spectral information due to significant advantages of spatial-spectral classification in terms of improved performance and accuracy [8]. In this regard, researchers have proposed certain techniques. First group of classification methods explores the spatial-spectral information before performing classification. For instance, using composite kernel methods [15–17], feature vectors [18] and morphological filters based feature extraction [19–22]. However, all these spatial features need human knowledge. Some HSI classification methods consist of incorporating spatial information into the classifier during the classification process. Similarly, some classification techniques try to include spatial

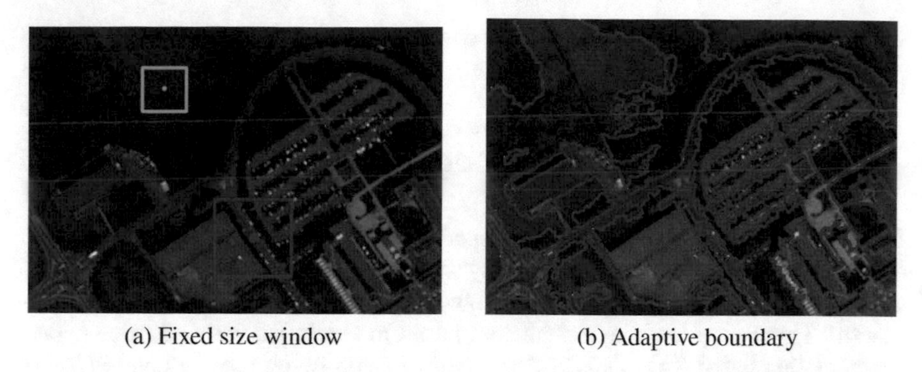
(a) Fixed size window (b) Adaptive boundary

Fig. 5.2 Spatial region selection by **a** fixed square window and **b** adaptive boundary adjustment based segmentation

information after the classification either by spatial regularization or by decision rule such as [23].

Although the aforementioned techniques provide promising classification results, the spatial contextual part in the remotes sensing domain is not adequately analyzed as the contour and size of the approved spatial area is not flexible. For instance, in the spatially homogeneous areas of HSI, large region sizes should be selected while heterogeneous areas require small region sizes. So the frame of the structure should be changed to the actual object region in the HSI as explained in Fig. 5.2.

In recent years, deep learning is successfully implemented in areas of image classification [24, 25], object detection [25, 26], language processing [27], and speech recognition [28]. Deep learning has also been implemented in HSI classification [2, 29]. However, fixed-size scanning window in deep learning is one of the major limitations as the size of the scanning window is fixed without any prior spectral-spatial knowledge, hence it can either have multiple classes or a subset of a single class within that window as shown in Fig. 5.2. It ultimately results in more complexity and affect the classification performance. Moreover, availability of the sufficient training samples is also limited in HSI, unlike typical RGB images, where huge labeled training data is available. In modern times, quite a few deep structures have been recommended for HSI processing [30]. Deep learning based Stacked Auto-Encoder (SAE) is implemented in [29]. Later, the enhanced version was proposed in [31]. Moreover, deep belief network (DBN) was also proposed in [32]. These techniques have proved useful in extracting robust features. However, these aforementioned models have certain limitations:

- Size of the scanning window during the learning process is fixed without any prior knowledge. Hence, scanning window size may enclose several classes or a subgroup of the same class.
- Due to fixed spatial window size, DL architectures are incapable of extracting the spatial features in HSI as static size window doesn't take any spatial information into account.

- Use of traditional window scanning technique for each target pixel as an input to DL with HSI data is very complex, due to the presence of hundreds of spectral channels.
- Because of the complete connection of different layers in DL architectures, they require plenty of sampled training labels to train several constraints, which is a crucial issue in HSI.

In this phase of book, we propose three spatially adaptive hyper-segmentation based deep learning architectures to efficiently utilize the spatial contextual data where traditional pixel-by-pixel scanning window is replaced by the structural/voxel information. Three deep learning architecture based methods, spatially adaptive hyper-segmentation based SAE, (SAHS-SAE) model, Deep Belief Network and PCANet are developed in this part of the book. Proposed methods utilize detailed available spectral channels when the existing training labeled data is not sufficient. Each resulted segmented area [33] of the scene can be reflected as a confined spatial structure whose dimensional and shape can be flexibly adapted for various object structures. Proposed methods first employ an efficient hyper-segmentation approach [33], designed in the previous chapter, to partition the remote sensing scene into various spatially associated structures. Formerly, it is expected that pixels in each segmented region also possess the same spectral signatures, and their correlations are exploited via above-mentioned DL architectures [29], where pixel scanning window is replaced by the adaptive spatial segmented region. As segments are of different size but DL takes fixed size as an input, therefore, only the optimum segment size is taken and each segment is resized accordingly. Proposed techniques exploit multiple-layer DL architectures to determine the abstract representation of HSI, respectively. Moreover, a new method for integrating spectral and spatial features is proposed. The main contributions of this phase of thesis can be summarized as follows:

- Traditional pixel-wise scanning window in DL is replaced by the object-level features which greatly reduce the complexity and leads to better classification accuracy.
- Spatial features are exploited through adaptive boundary adjustment where fixed-size window is replaced with adaptive hyper-segments which flexibly vary the size and shape conferring to the spatial structures in HSI which results in the maximization of feature differences between different class and minimization within classes.

The main contributions of this work focus on how to capture and utilize the spatial features and combine spectral deep features for HSI classification. The remaining work in this direction is presented as follows. In Sect. II, band selection and segmentation preprocessing steps are briefly summarized as the two steps are already discussed in detail in previous chapters. The next section describes the designed SAE-based HSI classification framework. Section 5.3 presents DBN while Sect. 5.4 describes PCANet-based classification of HSI. Each section includes experimental results with three real and diverse hyperspectral datasets and comparison with the state-of-the-art existing classification techniques, detailed discussion on parameters, and summary at the end.

5.1 Preprocessing

5.1.1 Band Selection

Hyperspectral images comprise hundreds of spectral bands. However, these bands also include a significant number of redundant and noisy bands, which can propagate adverse statistical and geometrical features that can cause inefficiency and inconvenience for HSI classification. Band selection (BS) techniques such as A^3BC [34] that effectively select the subset of most discriminative and informative spectral bands based on adaptive boundary adjustment based group-wise band-categorization framework without compromising the original content. Linear projection based principal component analysis (PCA) results in the loss of important spatial and structural information, hence affects the classification accuracy. Both PCA and band selection based HSI classification techniques were investigated in previous studies [35, 36]. Results proved the superiority of band selection based techniques over PCA as spatial structural information tend to exist in lower components rather than major components. Therefore, band selection technique (i.e., A^3BC) is used in this method.

A^3BC as presented in Chap. 3 is an efficient BS approach that extracts the most distinctive and informative bands based on the tri-factor weight model, which is evaluated based on the measurement of movement/adjustment of edges. First, HSI is divided into R clusters with cluster set, $\mathbf{C} = (C_1, \ldots, C_R)$. Cluster size W_{size}, cluster shift W_{size}, and cluster spectral-spatial contextual (SSC) W_{ssc} factors are calculated. Final band selection factor is then evaluated as

$$W = W_{size} \times W_{shift} \times W_{ssc} \tag{5.1}$$

More detailed implementation can be found in Chap. 3.

5.1.2 Hyper-Segmentation-Based Spatial Feature Extraction

In order to effectively segment the spatially similar structures in HSI, we segment the HSI using boundary movement/adjustment based technique proposed in our previous chapter which segments the regions using local structural regularity and self-similarity. The size and shape of each region is adaptively changed based on the tri-factor criteria. The detailed segmentation procedure is presented as follows:

1. Divide the initial hyperspectral M channel image into $\frac{M}{N}$ N-channel groups
2. For each group of channels, perform the following steps:
    a. Divide each channel group into m preliminary hexagonal-shaped structures/clusters.
    b. Repeat the subsequent steps till no further evolution is possible, i.e., no pixel going into a diverse segment.

- Compute the main class of each preliminary divided segment/cluster.
- Compute the gradient of existing edge pixels of each segment/cluster.
- Compute the straightness criteria for each cluster edge.
- Established on the calculation, compute the similarity measure, i.e., force of attraction/repulsion between each class centroid and its edge parts.
- Develop the edge, established on the similarity measure.

3. For each band group, weight to every edge pixel is assigned in accordance with the predefined criteria to acquire a real object boundary.
4. Merge N channels to obtain a weighted segmentation of HSI.

In previous work [29], the authors utilize the spatial information to enhance the performance. However, since the size and shape of the spatial window is static, the spatial information still may not be appropriately exploited. For instance, if the window size for the target pixel is selected too large [red region in Fig. 5.2a], some uncorrelated pixels might be included, thus results in classification performance deterioration. In contrast, too small window size in the smooth region [green region in Fig. 5.2a] results in insufficient exploitation of spatial information. On the other hand, each hyper-segmented region is intuitively a regular region, whose shape and size can be adaptively varied according to different spatial structures as described in Fig. 5.2b.

5.2 SAE Based Shape-Adaptive Deep Learning for Hyperspectral Image Classification

In this part of chapter, our focus is to exploit the contextual features through adaptive boundary based segmentation to enhance the SAE process where pixel-level scanning window is replaced by object level segments as an input to SAE. As segments are of different sizes but SAE takes fixed size as an input, therefore, only the optimum segment size is taken and each segment is resized accordingly. A general flowchart of the framework is presented in Fig. 5.3, which consists of three main phases: first, boundary adjustment based band selection approach is utilized to select the most discriminative and informative bands. Secondly, adaptive boundary movement based segment approach is employed to partition the scene into spatially similar regions. Thirdly, the resulted segmented boundaries are used as adaptive windows for computing SAE-based shallow and deep features. Next, learned spectral-spatial features are fed into multinomial logistic regression (MLR) for classification. A detailed description of each component is described in the remainder of this section.

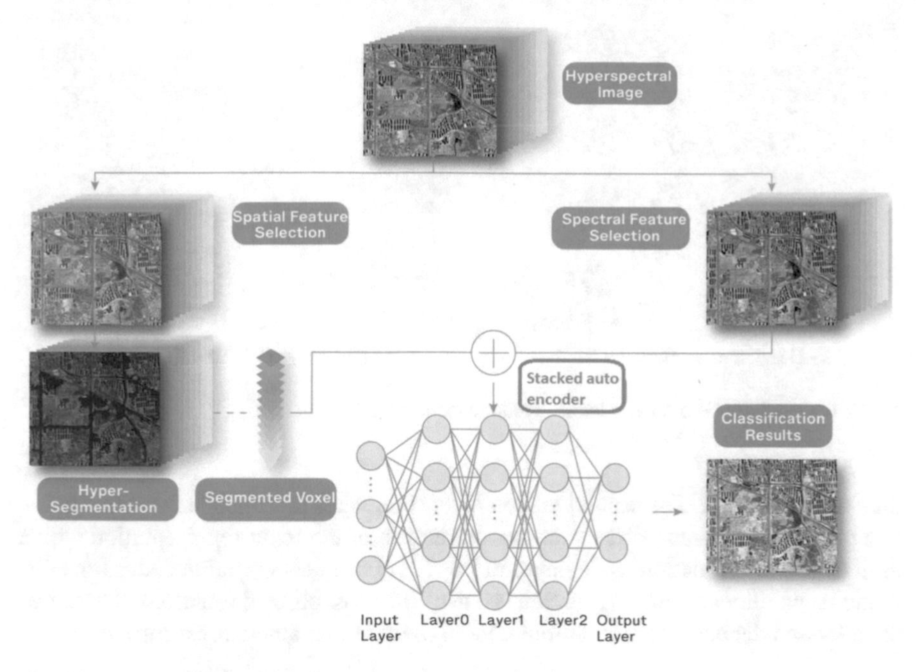

Fig. 5.3 Framework of the SAHS-SAE-based classification

5.2.1 Feature Extraction

In this part, we will have a deeper insight into the working of auto-encoders that will help you build a strong fundamental knowledge about the topic so let's move forward and look at it. One of the major goals for which auto-encoder was designed was to compress the data. Data compression is one of the big topics as we all know.

Auto-encoder is un-supervised deep learning architecture that perform regeneration of the representation that it receives as an input with compression. The purpose of an auto-encoder is to acquire a compressed distributed depiction for the given input, typically for the purpose of dimensionality reduction and feature extraction. However, why do we need auto-encoders when we have PCA for this purpose, i.e., for dimensionality reduction. An auto-encoder can learn nonlinear transformations contrary to PCA through a nonlinear function with the aid of several neural network layers. It performs the learning procedure through convolutional layers. Convolutional layers are considered virtuous for the images, videos, and time series data. Learning several parameters with convolutional layers is more effective and efficient. AE can play a very major role in determining the deep and variant features of the high-dimensional representation without requiring the training samples [25]. In particular, the raw high-dimensional data is fed into the first convolutional layer of auto-encoder, which transforms the input data x to a new encoded representation with an encoder function. This new transformed first layer can be taken as a new encoded

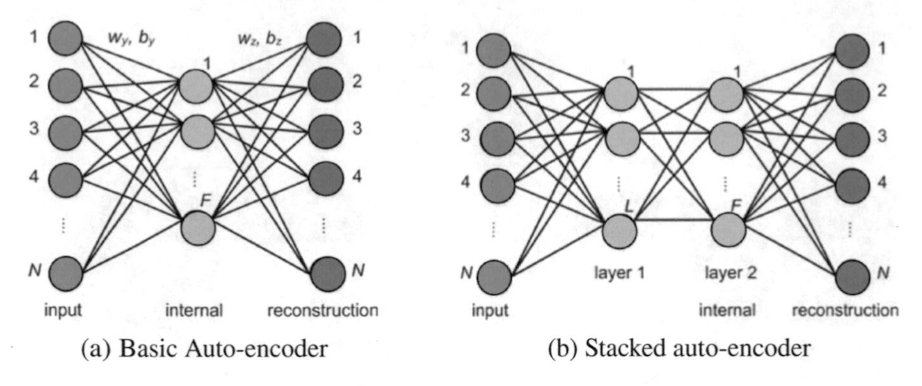

(a) Basic Auto-encoder (b) Stacked auto-encoder

Fig. 5.4 General SAE model. It learns a hidden feature from input

data representation. The second hidden layer then takes this encoded representation and regenerates or reconstructs an approximation of the input representation with a help of decoder function as depicted in Fig. 5.4. For encoder and decoder transformations, Nonlinear sigmoid function is one of the most utilized functions. These two encoder and decoder representations can be defined in mathematical form as

$$a = f(w_a x + b_a), b = f(w_b y + b_b) \tag{5.2}$$

where **a** is calculated from input data x through weights w_a and bias b_a. The aim is to train the auto-encoder in such a way to minimize the error between x and z by calculating the improved quantities of parameters

$$\underset{w_a, w_b}{\arg\min}[error(x, b)] \tag{5.3}$$

Stacked Auto-Encoder (SAE) is the extension of auto-encoder procedure. AEs are linked on over the other to transform into SAE. SAE has developed itself as one of the prevailing tool for effective extraction of deep features [29]. One of the cost functions that is employed for SAE is cross-entropy [29]. While in conjunction with the sigmoid activation function for the SAE is provided as

$$c = -\frac{1}{p}\sum_{a=1}^{p}\sum_{q=1}^{d}[x_{aq}\log(z_{aq}) + (1 - x_{aq})\log(1 - z_{aq})] \tag{5.4}$$

where d and p represent the vector size and mini-batch size. Small-patch stochastic gradient descent is employed to optimize this equation [29]. Let (X, Y) represents the training set, where x^n is the corresponding label for y^n. The complete process can be briefed as follows:

1. Each training sample x^n is encoded by the encoder $z^n = f(x^n)$, where $f(.)$ is the encoding function.
2. Decode z^n to the reconstruction by $h^n = g(z^n)$, where $g(.)$ is the decoding function.
3. Optimize the encoding and decoding functions parameters by reducing the error between the reconstructions and the inputs over the whole training set.

To monitor the learned spectral-spatial features from SAE hidden layers into different classes, we add MLR as an output layer.

5.2.2 Experimental Results and Performance Comparisons

This section assesses the achievement of the designed technique on real airborne HSI datasets captured by different sensors. The proposed method is applied to diverse and complex datasets containing both urban and natural structures. We used three datasets from two different sensors. They are Indian Pine, Pavia University, and Salinas Valley. The detailed specification of each dataset is described in Chap. 2, Sect. 2.3.

5.2.2.1 Parameter Setting

The performance of the hyper-segmentation based stacked auto-encoder SAHS-SAE method is compared with the well known and widely used HSI classification techniques. These methods include Stacked Auto-Encoder with Logistic Regression (SAE-LR) [29], augmented Lagrangian multilevel-logistic (LORSAL-MLL) [37], SVM with RBF kernel (SVM-RBF), simultaneous orthogonal matching pursuit (SOMP) [38], (MPM-LBP) [39] and CNN [40]. For every image, we split the available sample data into training and testing with the ratio 1:9. That is 10% as training and 90% as testing data. The training and testing samples are shown in the respective tables. Three widely used measurement matrices for HSI classification are utilized to analyze the performance of all the classification techniques used in this work. Overall accuracy (OA) denotes data that is classified accurately over test data. Average accuracy (AA) refers to the mean of the accuracy of each class. Kappa coefficient shows a robust measurement of the degree of agreement between the ground truth and the final classified map. Detailed discussion on each measurement matrix is presented in Chap. 2, Sect. 2.3.

Table 5.1 Parameter used for hyperspectral images

Dataset	Parameter	Value
Indian Pine	No of hidden layers	2
	No. of units in input layer	No. of selected bands of HSI
	No. of units in hidden Layer 0	180
	No. of units in hidden Layer 1	100
	No. of units in output layer	16
Pavia University	No. of hidden layers	2
	No. of units in input layer	No of selected bands of HSI
	No. of units in hidden Layer 0	180
	No. of units in hidden Layer 1	100
	No. of units in output layer	9
Salinas	No of hidden layers	2
	No. of units in input layer	No of selected bands of HSI
	No. of units in hidden Layer 0	200
	No. of units in hidden Layer 1	120
	No. of units in output layer	16

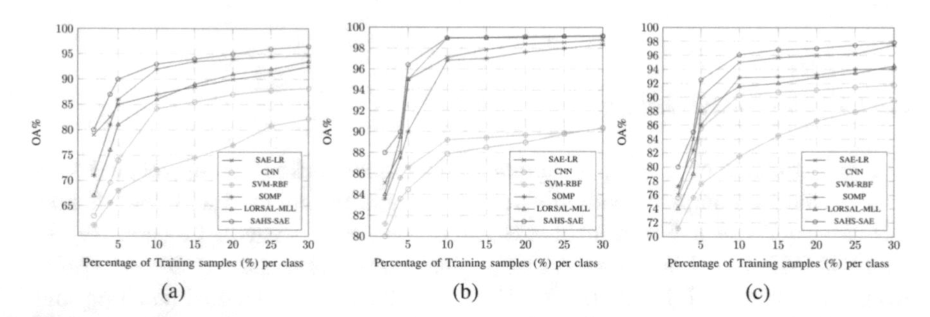

Fig. 5.5 Classification results with corresponding OA curve for **a** Indian Pine **b** Pavia University **c** Salinas datasets

5.2.2.2 Classification with Spectral-Spatial Features

We begin to exploit the potential of the proposed method by first extracting the spectral and spatial features and then classifying them using MLR. Parameters used after running a number of experiments for each dataset are described in Table 5.1.

Indian Pine dataset is known for high complexity in terms of classification for low spatial resolution. Results confirm that spatial learning based feature representation has a great impact because spatial/structural features can considerably prevent salt and paper noise and give the feature learning algorithm more discriminating power to effectively classify the pixel. Visual classification results are shown in Fig. 5.6. As can be observed, other techniques that consider spatial information with fixed-size windows results in noisy estimations in the map. These approaches also don't result

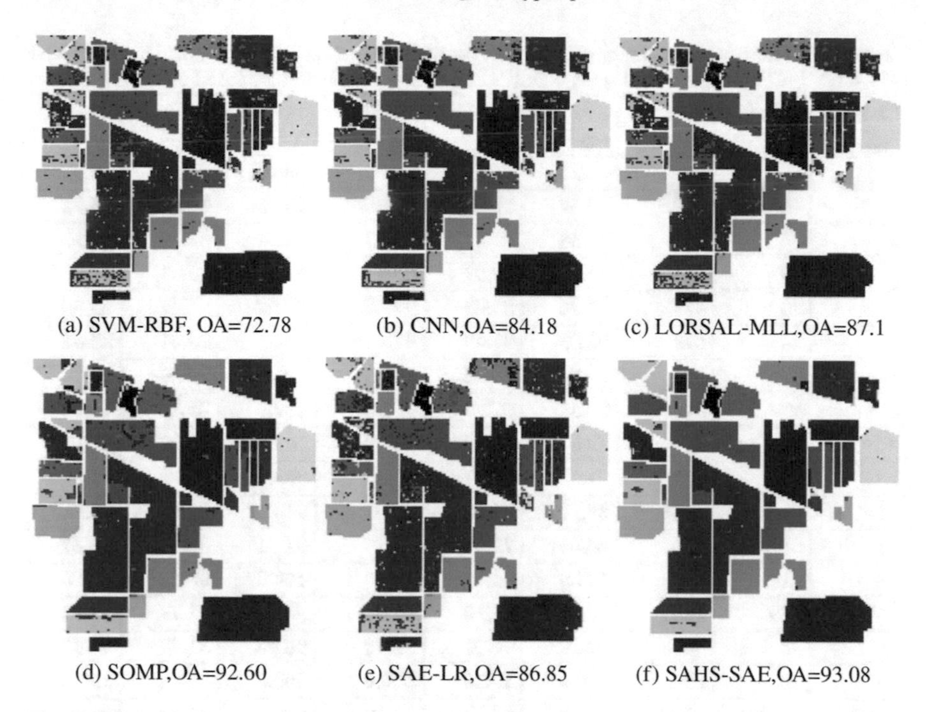

Fig. 5.6 Classification results with corresponding OA for Indian Pine dataset with 10% training samples

in accurate classification in detailed and near-edge regions. On the other hand, the proposed SAHS-SAE approach not only provides accurate estimations in the detailed area but also achieves a smoother appearance. Table 5.2 shows the quantitative results of the proposed technique evaluated with different well-known existing techniques. From the table, it can be noticed that the designed technique performs best on this challenging dataset with OA, AA, and k. We have also implemented experiments to evaluate the response of varying the quantity of training data per class. By keeping all the other parameters constant, we changed the given number of labeled data per class from 10 to 30%. Figure 5.5 shows the accuracies of all the methods where OA is plotted against various percentages of training samples. It can be noticed that the accuracy increases with the increase in the training samples. But OA tends to be stable after 10% training samples. The proposed technique clearly outperforms all other methods and more importantly, it performs even better with less training data.

Second experiment was implemented on Pavia University scene, which consists of multiple structures of varying sizes. Classification results of each class are presented in Table 5.3 while visual results are presented in Fig. 5.7. As it can be observed that the proposed technique outperforms the other existing techniques both visually and quantitatively.

Salinas dataset comparatively consists of large spatial structures. The experimental results of Salinas dataset along with the comparison with existing techniques are

Table 5.2 Classification accuracy of each class for the Indian Pine dataset obtained by SVM-RBF [41], CNN [40], LORSAL-MLL [37], SOMP [38], SAE-LR [29] and proposed SAHS-SAE

Class	Training	Test	SVM-RBF	CNN	LORSAL-MLL	SOMP	SAE-LR	SAHS-SAE
1	5	49	60.77	56.79	**100.0**	**100.0**	93.33	98.81
2	143	1291	77.68	52.17	87.46	95.05	84.66	**95.42**
3	83	751	79.35	85.33	81.23	96.24	84.39	**96.70**
4	23	211	91.05	87.92	88.41	97.93	73.08	**98.05**
5	50	447	84.36	85.22	97.49	**98.68**	93.47	95.54
6	75	672	92.03	97.49	97.34	98.36	93.41	**98.61**
7	3	23	69.61	74.62	**100.0**	**100.0**	**100.0**	**100.0**
8	49	440	59.31	67.99	97.98	**100.0**	95.11	**100.0**
9	2	18	79.61	58.87	**100.0**	**100.0**	**100.0**	**100.0**
10	97	871	97.53	98.77	83.64	95.44	85.78	**95.73**
11	247	2221	85.21	87.62	86.31	91.80	83.46	**92.11**
12	61	553	63.64	72.42	91.79	91.71	81.62	**92.27**
13	21	191	**100.0**	93.33	**100.0**	**100.0**	98.52	**100.0**
14	129	1165	87.18	71.79	97.04	**98.0**	91.77	95.45
15	38	342	90.19	90.91	86.18	96.07	81.79	**96.95**
16	10	85	**100.0**	**100.0**	**100.0**	**100.0**	98.88	**100.0**
Overall accuracy			72.78	84.18	87.18	92.60	86.85	**93.08**
Average accuracy			82.39	80.08	90.83	94.14	89.95	**95.09**
Kappa coefficient			0.6931	0.6852	0.8536	0.9155	0.8495	**0.9275**

Table 5.3 Classification accuracy of each class for the Pavia University dataset obtained by SVM-RBF [41], CNN [40], LORSAL-MLL [37], SOMP [38], SAE-LR [29], and the proposed SAHS-SAE

Class	Training	Test	SVM-RBF	CNN	LORSAL-MLL	SOMP	SAE-LR	SAHS-SAE
1	597	6034	84.01	85.62	**100.0**	**100.0**	96.88	98.78
2	1081	16971	88.90	88.95	98.50	97.21	98.30	**99.01**
3	189	1910	87.57	80.15	**96.27**	92.47	91.09	94.50
4	276	2788	96.09	96.93	91.03	74.39	99.15	**99.40**
5	121	1224	99.91	99.30	98.45	93.36	99.85	**99.89**
6	453	4576	93.33	84.30	97.24	**99.54**	96.44	97.30
7	120	1210	93.98	92.39	**97.94**	96.76	94.12	96.96
8	331	3351	82.94	80.73	**99.94**	99.58	93.27	97.14
9	85	862	99.60	99.20	**100.0**	**100.0**	**100.0**	**100.0**
Overall accuracy			89.24	87.90	98.95	97.84	97.12	**98.98**
Average accuracy			91.82	89.73	97.28	93.96	96.56	**97.46**
Kappa coefficient			89.01	87.20	0.9819	0.9628	0.9615	**0.9875**

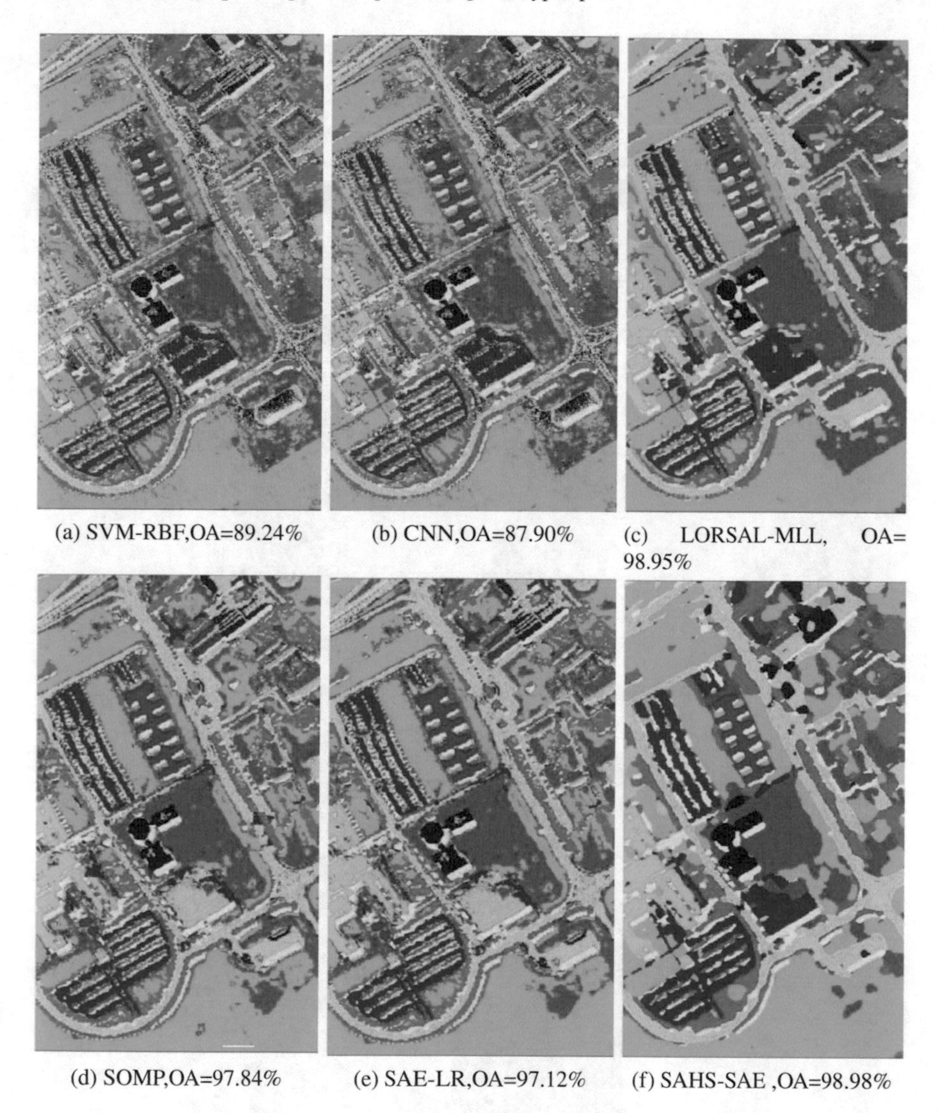

(a) SVM-RBF,OA=89.24% (b) CNN,OA=87.90% (c) LORSAL-MLL, OA= 98.95%

(d) SOMP,OA=97.84% (e) SAE-LR,OA=97.12% (f) SAHS-SAE ,OA=98.98%

Fig. 5.7 Classification results with corresponding OA for Pavia University dataset with 10% training samples

presented in Table 5.4. The performance of the proposed technique on each of the 16 classes, OA, AA, k along with the various existing techniques is described. Classification maps are presented in Fig. 5.8. From these results, it can be observed that the proposed technique clearly outperformed the presented existing techniques.

Experiments on these diverse datasets demonstrate that incorporating the spatial information and replacing the fixed window with segmented spatial voxel in SAE greatly improves the classification performance. Correlation among and within each

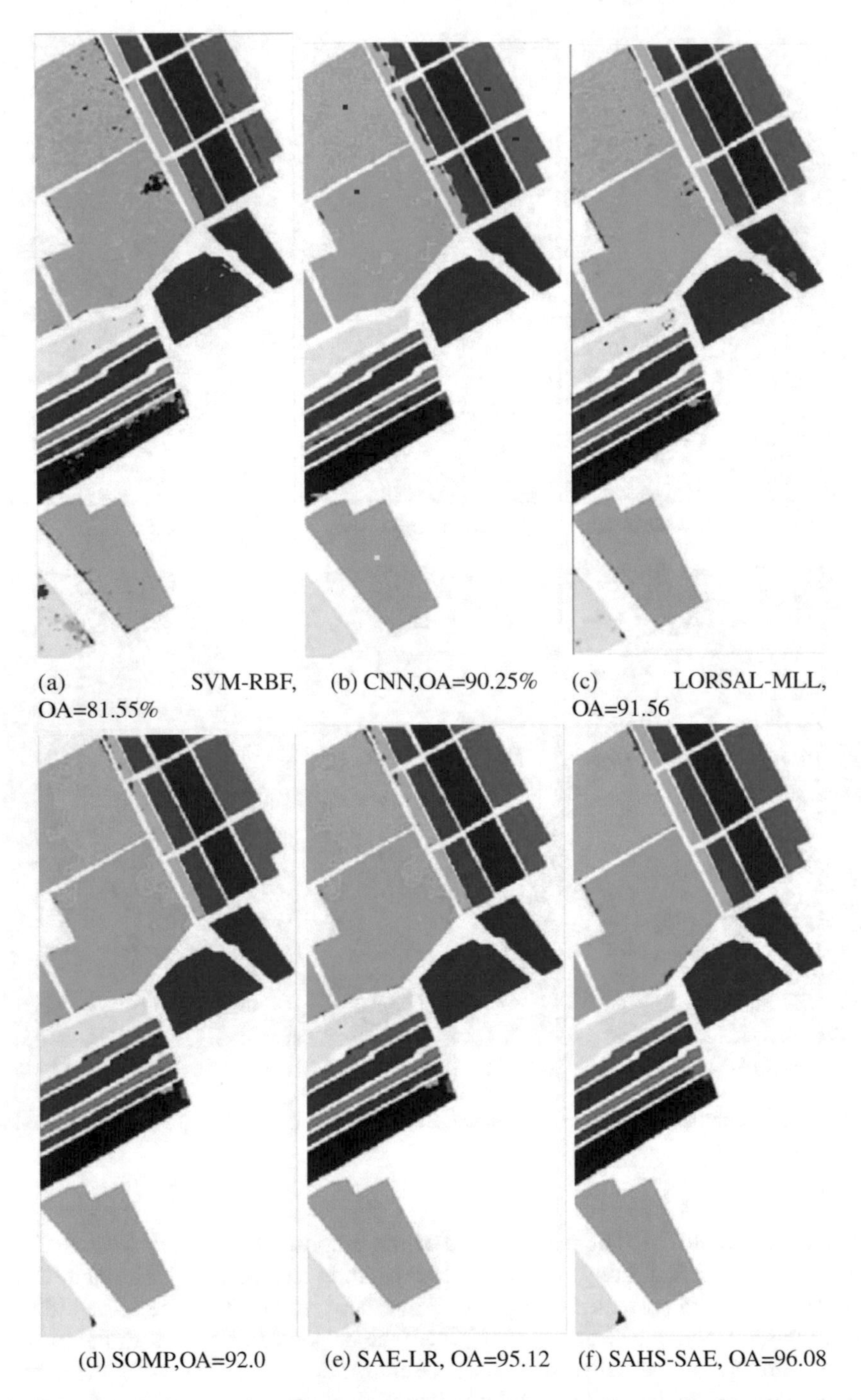

(a) SVM-RBF, OA=81.55% (b) CNN,OA=90.25% (c) LORSAL-MLL, OA=91.56

(d) SOMP,OA=92.0 (e) SAE-LR, OA=95.12 (f) SAHS-SAE, OA=96.08

Fig. 5.8 Classification results with corresponding OA with 10% training samples for Salinas dataset

Table 5.4 Classification accuracy of each class for the Salinas dataset obtained by SVM-RBF [41], CNN [40], LORSAL-MLL [37], SOMP [38], SAE-LR [29] and proposed SAHS-SAE

Class	Training	Test	SVM-RBF	CNN	LORSAL-MLL	SOMP	SAE-LR	SAHS-SAE
1	20	1989	96.81	98.37	**100.0**	99.90	**100.0**	**100.0**
2	37	3689	94.67	99.12	99.65	**99.97**	99.38	99.42
3	20	1956	90.27	96.08	95.56	92.77	92.61	**94.70**
4	14	1380	98.61	99.71	98.08	95.76	98.79	**98.95**
5	27	1651	94.82	97.04	96.82	99.49	97.86	**99.54**
6	40	3919	97.61	99.59	**100.0**	99.32	99.87	99.91
7	36	3543	99.24	99.33	99.94	**100.0**	**99.38**	**100.0**
8	113	11158	54.69	78.64	**80.10**	86.61	83.32	**94.01**
9	62	6141	98.32	98.04	98.94	98.81	99.37	**99.41**
10	33	3245	81.91	92.38	94.54	95.21	96.07	**97.73**
11	11	1057	90.57	99.14	89.83	**97.53**	95.86	96.11
12	19	1908	92.43	99.88	95.11	99.95	99.48	**99.57**
13	9	907	98.07	97.84	95.30	97.73	94.84	**97.87**
14	11	1059	90.39	96.17	96.45	95.0	98.37	**98.45**
15	73	7195	60.06	72.69	78.28	71.31	78.34	**92.95**
16	18	1789	90.87	98.59	99.20	**99.94**	97.85	99.05
Overall accuracy			81.55	90.25	91.56	92.0	95.12	**96.08**
Average accuracy			89.33	95.19	94.86	95.58	96.94	**97.92**
Kappa coefficient			0.8015	0.8995	0.9058	0.9456	0.9715	**0.9875**

spatial segmented voxel is exploited for feature extraction which greatly enhances the performance of SAE.

5.2.3 Summary of the Proposed Integration of Spectral-Spatial Information Method for Deep Learning Based HSI Classification

We have proposed a SAHS-SAE technique by analyzing a state-of-the-art SAE model and solving its concerns by construction. Despite its outstanding learning capability, pixel-wise scanning window might limit the capability of SAE in remote sensing context. We, therefore, proposed SAHS-SAE, a two-step approach that integrates the effect of spatial features and replaces the fixed pixel-wise scanning window with adaptive spatial structural window making it possible to fully exploit the capability of SAE. In this research, an innovative SAHS-SAE technique is proposed to exploit spatial contextual features via hyper-segmentation (HS) and incorporate it into multi-layer SAE for effective HSI classification. Unlike previous SAE-based feature extraction methods, the SAHS-SAE approach develops a shape-adaptive neighborhood structural area for each sample pixel instead of using a square window of fixed size. The size and shape of the structural region can be flexibly adapted according to the actual HSI spatial structures, which results in an effective and adaptive exploitation of spatial contexts. SAHS-SAE then uses the multi-layer SAE to effectively exploit the HS-based spatial features and selected spectral features within and among segmented regions. Furthermore, for efficient HSI classification, A^3BC-based feature selection approach is employed prior to feature extraction. Experimental results on three AVIRIS and ROSIS sensors based on diverse HSI datasets demonstrate that proposed SAHS-SAE approach produces better performance and outperform existing well known and widely used HSI classification approaches, especially in images with small spatial structures. Furthermore, SAHS-SAE as one of the robust feature extractors works well in heterogeneous regions, particularly for complex urban scenes.

One of our future research direction is to develop a more systematic method of adaptively selecting the number of hyper-segments based on different structural information. Moreover, there is a strong motivation to apply the SAHS-SAE approach to other HSI areas like variation detection, noise detection, and target recognition.

5.3 DBN-Based Shape-Adaptive Deep Learning for Hyperspectral Image Classification

In this part of chapter, we are primarily interested in the classification of HSI based on the Deep Belief Network. As discussed earlier, successful and exact hyperspectral images (HSI) investigation has received an increasingly significant impact with developments in remote detecting sensors. These sensors are now attaining useful information with several many spectral channels alongside the definite spatial data of the scene. The rich spectral and spatial data if effectively used can create higher order correctness.

HSI order has fundamentally centered around either Feature Selection (FS) or Feature Extraction (FE) as an approach to encourage classifier [3] for characterization. FS comprises selection of fitting subset of spectral groups, while FE includes the change of the information into a space of decreased dimensionality [2]. Support Vector Machines (SVMs) classifier performed well on high-dimensional information with hardly any preparation samples [3, 4]. Normally, feature extraction/dimensionality decrease methods, for example, Principal Component Analysis (PCA) and Independent Component Analysis (ICA) are applied, which brings about the loss of definite information unavoidably and consequently impact the characterization precision. Besides, these conventional FE procedures may likewise results in the loss of local structural information.

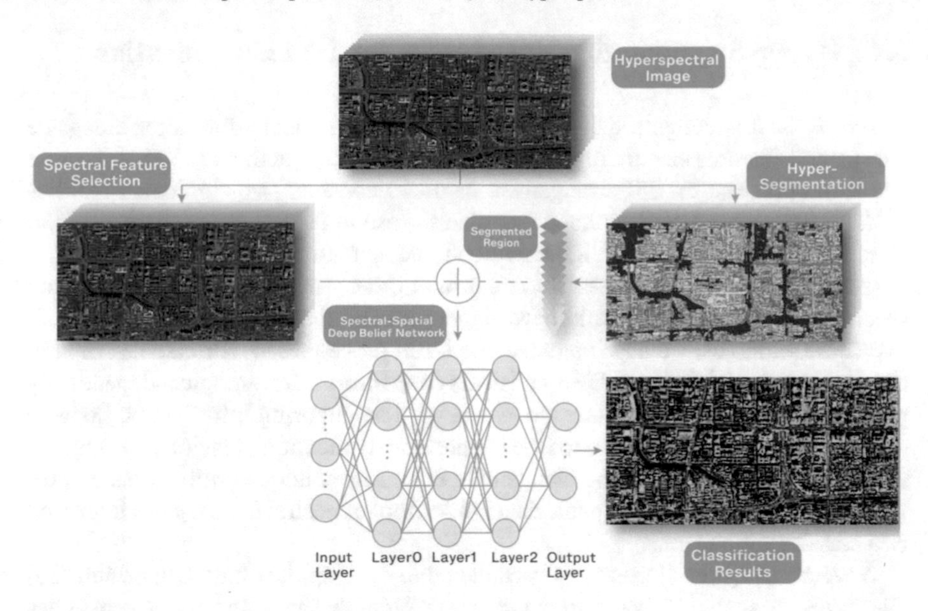

Fig. 5.9 Framework of the SDBN-based classification

In this part of the chapter, we are going to formulate hyper-division based Deep Belief Network (DBN) technique that adequately utilizes the spatial information of the HSI where pixel-by-pixel-based fixed examining window is replaced by adaptive local spatial information. Novel limit modification based hyper-division [33] has double impact. First, it keeps the spectral relationship in spectral channels and receives the article limits in the spatial area. Second, it unpretentiously settles and keeps the most informative groups and get rid of the non-related data and channels full of noise. It does so without losing the first hand vital channels that contain most of the information. The proposed system exploits the existing information with limited training information. The segments that are obtained as a result of applying [33], potentially contains a local spatial area with comparative spatial qualities in HSI. The weighted-division based spatial updated Deep Belief Network (SDBN) first endeavors a productive hyper-division approach to partition the HSI into expressive spatially comparable fragments. Pixels in each local segment are expected to have pixels with comparative spectral properties. Their connections are exploited through a deep belief network. The proposed procedure takes full advantage of multi-layer DBN to learn shallow and deep features of HSI (Fig. 5.9). The rest of the section is characterized as follows. The next subsection portrays the proposed system. Experimental analysis is discussed and analyzed in proceeding subsections followed by a summary.

5.4 Hyper-Segmentation Based DBN for HSI Classification

It is emphatically recognized that consolidating spatial logical information alongside the spectral features can enormously improve the order exactness [13, 14]. Feature extraction process for HSI arrangement ought to consider two significant realities [42]. (1) There is a high likelihood that pixels close in feature space share a similar class. (2) There is likewise a high likelihood that spatially neighboring pixels ought to likewise have a similar class. Putting it in another way, information with higher spectral/feature and spatial marks have bigger probabilities to have a similar class. To fuse these variables, we have upgraded the DBN procedure to extricate discriminative spectral-spatial features. DBN considers each information sample independently without thinking about any correspondence with neighboring information. So as to make the feature learning stage to fuse the previously mentioned factors, we upgrade the DBN process. It is done by maintaining the correspondence within data by providing DBN with contextual spatial data in addition to spectral data. It gives improved HSI classification results.

A weighted hyper-division is a versatile edge development based algorithm [33] that furnishes spatial relevant structures with adaptable shape and size that matches diverse real structures present in the HSI. In this part of chapter, SDBN framework enhances the DBN to practically utilize existing spectral–spatial information inside and among each resulted local regions. Normally, the proposed SDBN approach for the most part comprises of three stages as appeared in Fig. 5.9: (1) production of spatially versatile regions in HSI (2) feature selection (3) investigation of spectral–spatial features of these sectioned areas by means of DBN.

5.4.1 *Extraction of Spectral-Spatial Information of Spatial Segments via DBN*

A Deep Belief Network can be observed from a perspective of layers of neural network which is generative in nature. Each part consists of Restricted Boltzmann Machine (RBM). It comprises a layer for data input and a hidden layer. Hidden layer learn parameters to characterize data that detect higher order relationship between the received information shown in Fig. 5.10. The two RBM layers are linked through, a weights parameter w. RMB is independent of input–input layer of inner–inner layer connections. This restriction equip the inner units to be mutually independent. An integrated formulation of the energy function for inner unit h and input data units v is described as [32]:

$$F(y, z, \theta) = -\sum_{l=1}^{n} \frac{(y_l - b_l)^2}{2\sigma^2} - \sum_{k=1}^{m} a_k z_k - \sum_{l=1}^{n}\sum_{k=1}^{m} w_{lk} \frac{y_l}{\sigma_l} z_k \tag{5.5}$$

The probability distributions are given by

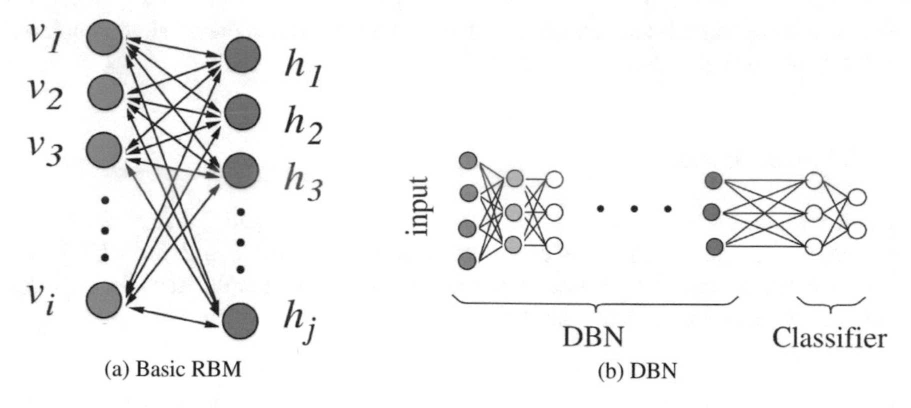

Fig. 5.10 General DBN model

$$P(h_j|v;\theta) = g\left(\sum_{i=1}^{n} w_{ij}v_i + a_j\right) \tag{5.6}$$

$$P(v_j|h;\theta) = N\left(\sum_{j=1}^{m} w_{ij}h_j\sigma_i^2 + b_i\right) \tag{5.7}$$

$N(.)$ RBMs are connected over one another and are trained in such a way as to get the optimized parameters to create DBN. The entire procedure can be explained in steps as

1. Given data is utilized to make the train the individual layer of DBN. In order to train the particular layer, Contrastive Divergence (CD) [43] is exploited.
2. Utilize the yield of the primary layer as a contribution for a second consecutive layer. Train the next layer in a similar manner.
3. Apply reiteration to 1 and 2 for a numerous number of layers as per requirement.
4. All resulted parameters should be fine tuned with the aid of available training data. Any classifier can be exploited for the given problem. We utilized Logistic Regression (LR) as a classifier.

In summarized form, first, we use hyper-segmented spatial region also, the spectral information after FS as an input. At that point, a DBN is applied to gain proficiency with the deep and unique features from the contributions through multi-layer DBN, at the end, LR is used to group and mark the pixels dependent on learning features.

5.4.2 Experimental Results and Performance Comparisons

So as to approve the presentation of the proposed strategy, we conduced investigation of three notable and testing genuine hyperspectral datasets procured through various sensors. These datasets have already been explained in detail in the previous chapters. Here, they are again described precisely

5.4.2.1 HSI Dataset

The three datasets incorporate Indian Pine, Pavia University and Houston University dataset. A detailed description of each dataset is described in Chap. 2, Sect. 2.3. So as to assess the grouping exactness, we randomly chose half samples each for training and testing purposes in all the datasets.

5.4.2.2 Parameter Setting

We directed tests on Windows 7 framework, with a 4.0 GHz processor having NVIDIA GeForce GTX 970. Theano was utilized as a coding apparatus. We picked a deep network with 2 layers for Indian Pine and Houston University dataset. We use deep neural network of depth 4 for Pavia University dataset as many researchers followed the same practice and received best results [32]. Each particular inner layer comprises 30 neurons for Indian Pine dataset. While for Pavia University dataset, 50 neurons for each inner layer are maintained. Similarly, 100 neurons were implemented for the Houston University dataset. As indicated by many researchers in this field, a number of inner neurons are more critical as compared to the number of inner layers as inner neurons play a more crucial role in feature extraction. The comparison of the results of the proposed technique is made with notable existing systems. These existing techniques consist of deep belief network with logistic regression (DBN-LR) [32] and augmented Lagrangian—multilevel logistic (LORSAL-MLL) [37].

5.4.2.3 Spectral-Spatial Classification Results

We utilized the initial 6 components of PCA and a window size of 7 by 7 for DBN-LR. A comparison of performance results of Indian Pine and Pavia University datasets and their correlation is presented in Table 5.5. While class-level precision for Houston University is portrayed in Table 5.6. Blended pixel is a significant problem in Indian Pine because of its low spatial size and little size. Results affirm that spectral-spatial classification utilizing logical feature extraction has a vital impact on accuracy performance. Since the neighboring data plays an important role in preventing the contribution of the typical present noises in an image. Figure 5.11 depicts the arrangement of three datasets.

Generally speaking, tested results show a huge improvement in HSI arrangement by consolidating spatial information and spectral feature selection. The framework has performed essentially well on the low spatial goals dataset.

Table 5.5 Classification accuracies for Indian Pine and Pavia University datasets

Dataset	Measurement	LORSAL-ML	DBN-LR	SDBN
Indian Pine	Overall accuracy (%)	87.18	95.95	**96.87**
	Average accuracy (%)	90.83	95.45	**96.75**
	Kappa coefficient (%)	0.85	0.9539	**0.9567**
Pavia University	Overall accuracy (%)	98.95	99.05	**99.18**
	Average accuracy (%)	97.28	98.48	**98.75**
	Kappa coefficient (%)	0.9819	0.9875	**0.9895**

5.4.3 Summary of the Proposed DBN-Based Shape-Adaptive Deep Learning Method for Hyperspectral Image Classification

In this part of a chapter, another new SDBN approach is designed with spatial relevant data exploitation by means of hyper-segmentation for powerful HSI classification. The size and state of each hyper-segment (HS) can be deftly adjusted as indicated by the genuine HSI spatial structures, which results in powerful exploitation of spatial contextual information for further use. SDBN at that point utilizes the multi-layer DBN to viably explore the HS-based spatial context and spectral characteristics both within and among sectioned hyper-segments. Exploratory classification performance shows that the proposed approach creates better execution than other existing methodologies in HSI classification. Particularly in images with very little spatial structures. Besides, SDBN as one of the vital feature extractor functions really well in heterogeneous data as well, especially for complex urban scenes. One of our future research heading is to build up a more precise technique for adaptively choosing the quantity of hyper-regions dependent on various spatial/basic data. Besides, there is a solid inspiration to apply the hyper-fragment based DBN to deal with other HSI applications, for example, change detection, noise detection, and object recognition.

5.5 PCANet-Based Boundary-Adaptive Deep Learning for Hyperspectral Image Classification

This section of the chapter presents another novel technique for HSI classification. The classification of Hyperspectral Image (HSI) is of primary importance in remote sensing image analysis. Due to the recent advances in the spectral and spatial resolution of remote sensors, many paths toward further improvements have been paved. The applications have been widened. Many applications now come under its umbrella including underground mineral discovery, city development planning, environmental management, and surveillance. But the involvement of these new applications comes

Table 5.6 Classification accuracy (%) of each class for the Houston University dataset by DBN-LR [32], LORSAL-MLL [37] and proposed method SDBN

Class	Training	Test	LORSAL-MLL	DBN-LR	SDBN
1	626	627	97.68	**99.20**	99.0
2	627	627	96.97	**99.60**	**99.20**
3	348	349	99.97	**100.0**	**100**
4	622	622	97.59	**99.60**	**99.60**
5	621	621	**99.95**	99.60	99.60
6	162	163	**98.15**	97.2	98.05
7	634	634	**98.42**	97.0	98.16
8	622	622	88.18	97.8	**98.0**
9	626	626	90.42	94.0	**95.25**
10	613	614	90.79	97.4	**97.95**
11	617	618	91.26	97.3	**98.1**
12	616	617	91.32	95.2	**96.26**
13	234	235	81.45	88.0	**91.1**
14	214	214	99.86	**100**	**100**
15	330	330	**100**	**100**	**100**
Overall accuracy			94.64	97.70	**98.98**
Average accuracy			94.82	97.50	**98.46**
Kappa coefficient			0.942	0.975	**0.9875**

at a cost. Not only do the meticulously captured details now incorporate complex spatial structures but also spectral regions that arise from inter-pixel relations. This reason among others makes the classification task more daunting. In recent times, newer methods have been created that exploit different classifiers and spectral-spatial information.

Following the success of deeper architectures in Artificial Intelligence, they have been developed for HSI classification as well. The improved accuracy and consequently performance can be attributed to sequential training by minimizing the MSE of all the data points extracted from various classes. For the training, SAE is used to pick up the deep features. Along with SAE, Deep Belief Network (DBN) and its variations are also being used that have significantly improved the accuracy. On top of that, Convolutional Neural Networks (CNN) are also used for the task of classification. But when training deep networks many parameters need to be learnt, and when the training data is limited it turns into a difficult task and hence serves as a bottleneck for HSI classification as data is scarce. Another limitation is the high demand for processing power and resources for learning deeper networks. But even with all these constraints, the spatial contextual information gets lost in the deeper architectures.

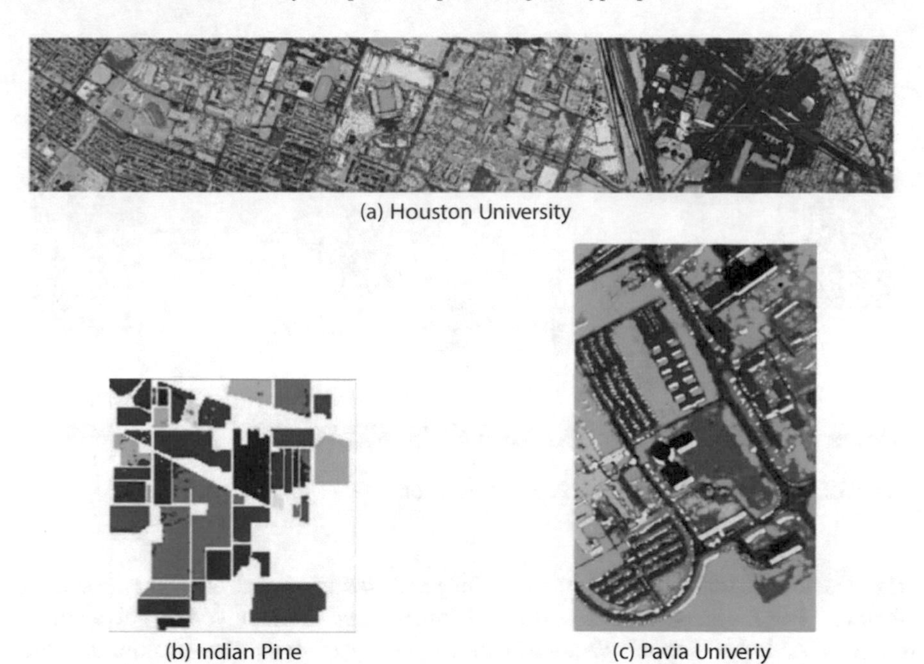

Fig. 5.11 Classification results using proposed method

Experiments have shown that by using spatial knowledge the classification performance can be improved. In this section, a Spatial-Adaptive Network (SANet) is presented that makes use of deep learning techniques and uses spatial and spectral features. The benefit is that in spite of the limited data for training, the HSI classification accuracy does not get affected. The proposed approach is inspired by Principal Component Analysis (PCA) Network (PCANet) [44]. PCANet performs well for normal RGB images [44]. There are three techniques that are used in this section. First one is adaptation of boundary based on hyper-segmentation on the basis of pixels that share the same spatial context. Second is dealing with high dimensionality by making use of extracted features. Third is making use of the spatial and spectral features that were extracted through PCANet. The remainder of this section is structured like first the detailed architecture is presented then the classification technique followed by the results and conclusion.

5.5.1 SANet-Based Spectral-Spatial Classification Network

There are two concurrent steps in SANet-based HSI classification. In the first step, feature selection approach is employed to get only the most contributing bands and thus minimize the complexity. Then adaptive boundary adjustment is used and the

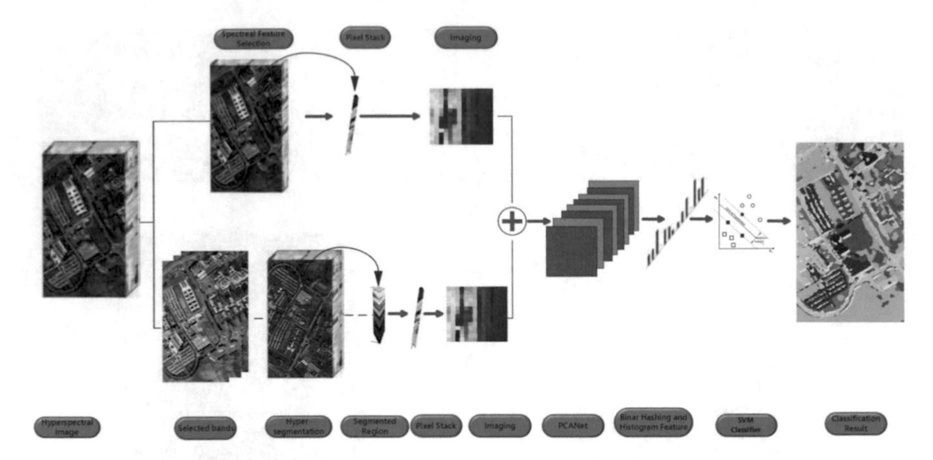

Fig. 5.12 Framework of the SANet-based classification

result is merged with spectral signature along with being vectorized to get the spatial features. For the next step, the extracted information is employed to get features using PCANet and Support Vector Machines (SVM) to get the classification results. Figure 5.12 shows the whole process. Now PCANet will be explained and then its modifications to fit the task of HSI classification will be highlighted; then the contextual features that were extracted are used as input to PCANet for increasing the HSI performance.

5.5.1.1 Spatial Feature Extraction Through Adaptive Hyper-Segmentation

Feature extraction is a very important step. The feature should be extracted in such a way to make sure that two points are given importance. The first one is that the probability of pixels belonging to the same class increases when the pixels share a common feature space. Another point is that the more the pixels are closer together in spatial space, the more is the probability of them belonging to the same class. When these two points are used in parallel, the discriminating power of PCANet gets improved. One bottleneck is that each pixel is considered separately in PCANet, i.e., without focusing on the neighboring pixels. To overcome this bottleneck, spatially similar pixels are chosen that share strong relationships in spatial features by employing hyper-segmentation. Hyper-segmentation works by grouping pixels that have the same spatial features as the neighboring pixels. Iteratively the dimensional of segments are changed till the true structural boundary can be identified. The segments that are obtained this way fulfill the criteria that are required. Segmentation is done by dividing the image into N segments each of size R and the grid being S. The center of the segment is calculated as the mean of the segment. The dominant features in all the segments are obtained and the gradients of the boundary pixel and straightness factor are all calculated. The similarity of each segment is calculated following the below equation and then the boundary is improved from this similarity measure. This iteration is done till no further changes take place. For the band selection, a certain approach is used that makes use of informative bands and the discrimination information is preserved.

5.5.1.2 PCANet

For a single image classification, PCANet is being used these days [44]. PCANet serves as an alternative to CNNs, i.e., to mitigate the complexity associated with them. One of the pivotal changes that have been made in PCANet is the shifting of convolution filters to PCA filters that are applied between layers. In this way, the ever so complex technique of optimization gets tackled. To further reduce this problem, binary quantization (hashing), histogram features, and linear SVM are all employed to address the issue of CNNs and deeper networks in general requiring complex optimization. PCANet primarily has three parts—filters, hashing, and histogram features. Further explanation of the conceptual parameters can be seen in [44].

A layered network is built using the PCA filters. All the vectorized patches are first accumulated in PCANet. Their size being $S_1 \times S_2$ with respect to each pixel and image I to construct a matrix Y. This matrix is formed for each and every image and then the results are combined into one.

$$\mathbf{Y} = [\mathbf{y_1}, \mathbf{y_2}, \ldots, \mathbf{y_N}] \tag{5.8}$$

N comprises of the whole pixels. The resulted error could be made minimum with

$$\underset{V \in R^{(s_1, s_2) \times M_1}}{\arg\min} \left[\mathbf{Y} - \mathbf{V}\mathbf{V}^{\mathbf{T}}\mathbf{Y}\right]_F^2, s.t. V^T V = E_{M_1} \tag{5.9}$$

where E comprises identity matrix of size $M_1 \times M_1$, V comprises principal eigen vectors of YY^T, and M_1 represents the number of principal eigen vectors. Filter can be expressed as

$$\mathbf{W_m^1} \doteq mat_{(s_1 s_2)} \left(q_m(\mathbf{Y}\mathbf{Y}^{\mathbf{T}})\right) m = 1, 2, \ldots, M_1 \tag{5.10}$$

where M_1 represents the no. of filters in a input layer and $mat(.)$ changes a vector into a matrix. The first PCANet layer can be captured as

$$\mathbf{I_i^m} \doteq \mathbf{I_i} * \mathbf{W_m^1}, i = 1, 2, \ldots, N. \tag{5.11}$$

The second layer of PCANet is obtained the same way. $M_1 M_2$ number of images are obtained while the number of filters are M_2. In the same way, multiple layers can be constructed of the PCA network. The authors suggest that four layers are enough for all mainstream tasks.

When the network is formed, binary quantization is used. Each image is split into blocks and histograms are made. Then the histograms are stacked together to form a vector. This vector is the feature expression.

5.5.1.3 Spectral-Spatial Classification

PCANet is essentially used to classify images. The network is altered by using the concatenated vector from the previous step. That vector is reshaped and molded into the form of an image using a data imaging process. After that, SVM is used to classify the features obtained. Suppose Z_i depicts the decisive stacked spectral-spatial vector. Z can be expressed in imaging terms as

$$z_i \xrightarrow{reshape} Z_i^{Spec} \tag{5.12}$$

In the end, in order to employ the classification process for the extracted features, an SVM classifier is deployed.

5.5.2 Experimental Analysis and Performance Comparisons

Two datasets are used to evaluate the classification accuracy of the method given above. To do that, two different datasets of images are used: Indian Pine and Pavia University datasets. The choice of these datasets is backed by the fact that many techniques that have been proposed have been checked against these datasets [39]. The datasets are challenging not only because of the rural and urban settings in which the images are captured but also because of the artificial and natural structures in them. Both the datasets are further described in Chap. 2, Sect. 2.3.

5.5.2.1 Indian Pine: AVIRIS Dataset

Through Airborne Visible Infrared Imaging Spectrometer sensor, Indian Pine dataset was attained over Northwestern Indiana. Altogether, it has a ground resolution of 17 m with a structural size of 145 $\times$ 145. There were 224 bands and 24 water absorption bands nearby 1400 and 1900 nm out of which 200 spectral channels remained. The ground truth map in Fig. 5.13 depicts 16 diverse crops based agricultural classifications. It also depicts the false color arrangement and ground truth. Spatial resolution, minute organization of different objects, and existence of varied pixels makes this dataset challenging.

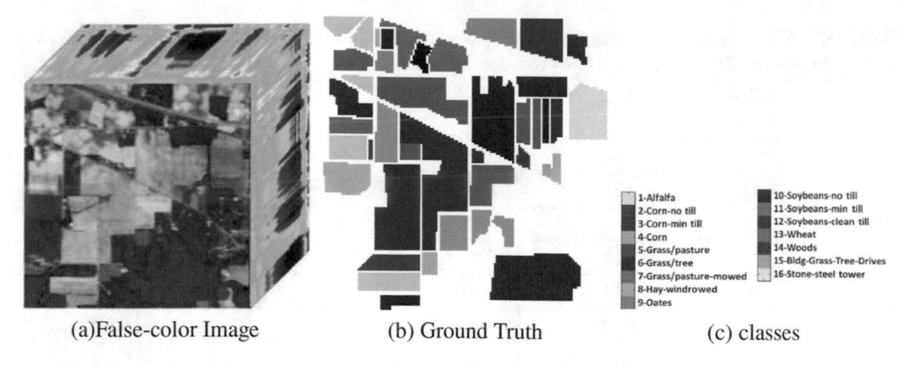

Fig. 5.13 Indian Pine dataset with 16 classes

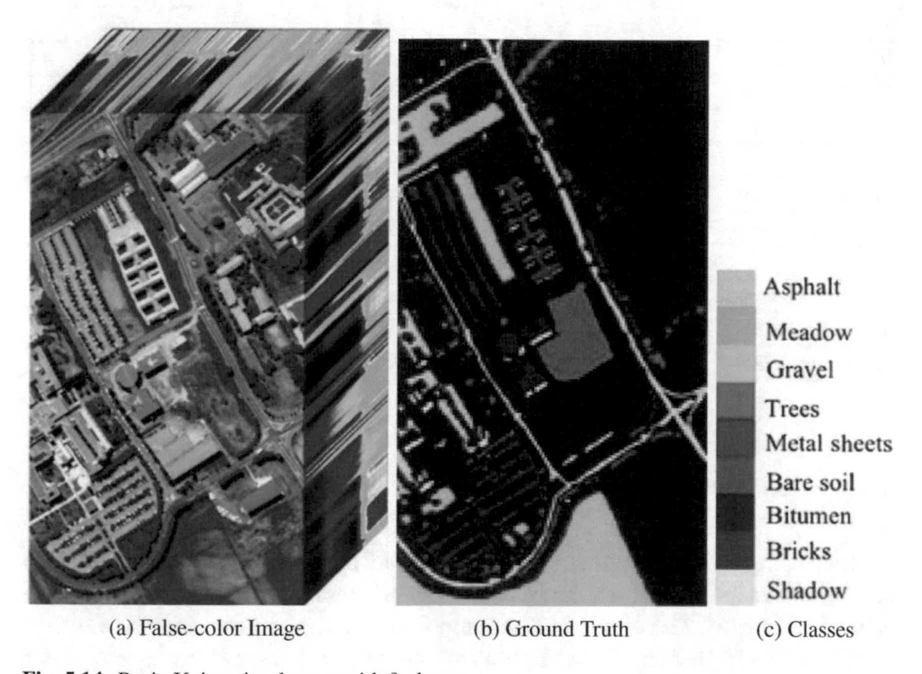

Fig. 5.14 Pavia University dataset with 9 classes

5.5.2.2 Pavia University: ROSIS Dataset

Pavia University dataset was gathered over Pavia University, Italy by Reflective Optics System Imaging Spectrometer sensor. Its spatial size is 610 rows and 340 columns of the image and 103 spectral channels. Practically, 9 structural classes are considered including mostly synthetic structures. False color composition and ground truth are depicted in Fig. 5.14.

Table 5.7 Classification accuracy of proposed SANet for each class for the Indian Pine dataset and its comparison with CNN [40], RNN [46], and SAE-LR [29]

Class	Training	Test	CNN	RNN	SAE-LR	Proposed SANet
1	5	49	56.79	70.59	93.33	**94.41**
2	143	1291	52.17	70.28	84.66	**91.15**
3	83	751	85.33	81.52	84.39	**92.70**
4	23	211	87.92	90.16	73.08	**94.05**
5	50	447	85.22	91.97	93.47	**95.54**
6	75	672	97.49	96.13	93.41	**98.61**
7	3	23	74.62	84.75	**100.0**	96.95
8	49	440	67.99	59.64	**95.11**	93.21
9	2	18	58.87	86.17	**100.0**	**100.0**
10	97	871	98.77	**99.38**	85.78	95.73
11	247	2221	87.62	84.97	83.46	**90.15**
12	61	553	72.42	77.58	81.62	**90.27**
13	21	191	93.33	95.56	98.52	**99.15**
14	129	1165	71.79	84.62	91.77	**95.45**
15	38	342	90.91	90.91	81.79	**91.95**
16	10	85	**100.0**	**100.0**	98.88	99.01
Overall accuracy			84.18	88.63	86.85	**91.48**
Average accuracy			80.08	85.26	89.95	**93.09**
Kappa coefficient			0.6852	0.7366	0.8495	**0.9075**

5.5.3 *Parameter Setting*

For a balancing ratio of 1:9 of each class, the selected data is arbitrarily positioned into training and testing. However, the PCANet network parameters setting are organized inspired from [44]. Therefore, SANet comprises $1 \times Input$ layer and $1 \times Output$ layer. Moreover, $2\times$ convolution layers are maintained. Each hidden layer consists of patch size of 7×7, including an overlapping ration of 0.8. These minute changes in the parameter are not much of a hindrance in the classification outcome [44]. For testing the classification performance, the following criteria is adopted:

1. *Overall Accuracy (OA)*: OA depicts the correct classification of the number of image spatial positions over the entire quantity of test trials taken.
2. *Average Accuracy (AA)*: AA measures the mean of the classification efficiency for the entire classes.
3. *Kappa Coefficient*: It maps the consent among the eventual efficiency result and the real ground truth map. Measurements of this coefficient are accurate as it considers the agreement occurring by chance [45].

Table 5.8 Classification accuracy of the proposed SANet for each class for the Pavia University dataset and its comparison with CNN [40], RNN [46], and SAE-LR [29]

Class	Training	Test	CNN	RNN	SAE-LR	Proposed SANet
1	597	6034	85.62	84.45	96.88	**98.18**
2	1081	16971	88.95	85.24	**98.30**	98.01
3	189	1910	80.15	54.31	91.09	**92.50**
4	276	2788	96.93	95.17	**99.15**	98.40
5	121	1224	99.30	99.93	99.85	**99.89**
6	453	4576	84.30	80.99	96.44	**97.01**
7	120	1210	92.39	88.35	94.12	**96.16**
8	331	3351	80.73	88.62	93.27	**95.14**
9	85	862	99.20	99.89	**100.0**	**100.0**
Overall accuracy			87.90	88.85	97.12	**98.01**
Average accuracy			89.73	86.33	96.56	**96.75**
Kappa Coefficient			0.872	0.8048	0.9615	**0.9675**

5.5.3.1 Compared Techniques

The deep learning architecture is compared with the state-of-the-art existing techniques like Stacked-Auto-Encoder (SAE-LR) [29], Convolution Neural Network (CNN) [40], and Recurrent Neural Network (RNN) [46].

5.5.3.2 SANet-Based Spectral-Spatial Classification Results

Here the correctness of SANet is examined for extracting the detailed features present in the HSI image and then further classifying them into their respective classes. Tables 5.7 and 5.8 show the achievements of OA, AA, and Kappa. These evaluation measures are evaluated on two available datasets for the developed SANet-focused feature extraction framework. This approach gives fair results and best parameters by utilizing 10. In Pavia University HSI image, approaches like SAE-LR and SANet have similar outcomes but SANet has the advantage of a larger dataset as compared to SAE-LR. Thus bigger dataset leads to improved performance. The low spatial resolution and mixed pixels make Indian Pine dataset more challenging. Table 5.7 shows the quantitative outcomes of each class for the Indian Pine dataset with other methods. It also highlights the great effect of spectral-spatial feature extraction technique on the classification accuracies. Thus it performs better than techniques in OA, AA, and k and visual classification result for each class for Indian Pine and Pavia University as shown in Fig. 5.15.

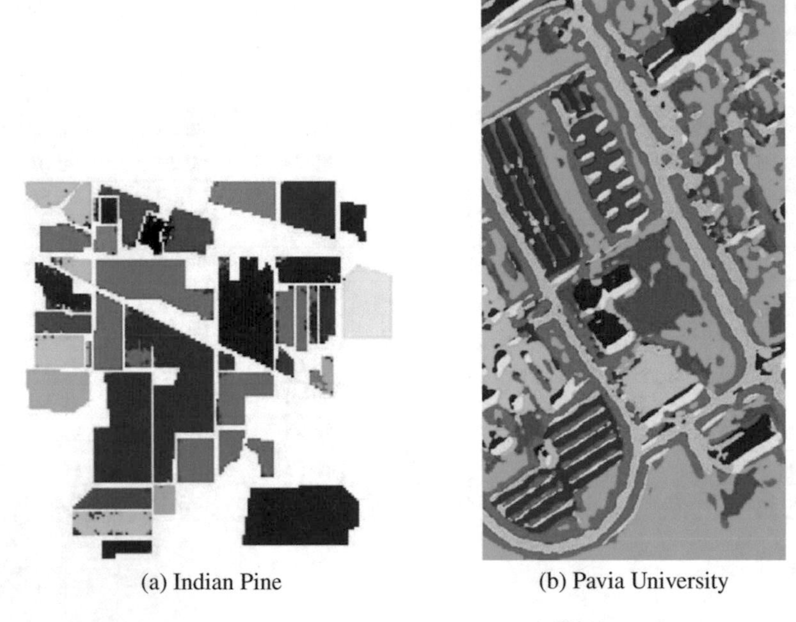

(a) Indian Pine (b) Pavia University

Fig. 5.15 Classification results of Indian Pine and Pavia University dataset

5.5.4 *Summary of the Proposed PCANet-Based Boundary-Adaptive Deep Learning Method for Hyperspectral Image Classification*

This work introduced a new simplified deep learning SANet technique for HSI classification by analyzing a state-of-the-art DL model and addressing its concerns by construction. Despite its outstanding learning capability, complex layered architecture and overlooking the spatial information limit the capability of deep learning in remote sensing context. We, therefore, proposed SANet, a simplified deep network based architecture that replaces inner-layer convolution filter banks with PCA filters hence reduces the complexity by avoiding the optimizing process. Moreover, spatial information is exploited through hyper-segmentation and integrated with spectral characteristics. SANet then utilizes the multi-layer deep learning network to effectively exploit the hyper-segmentation-based spatial features and selected spectral features within and among segmented regions. Experimental results on two AVIRIS and ROSIS sensors based challenging datasets reveal an improved classification performance especially in small structural regions. One of the future tracks is to reduce the training samples and further improving the performance.

5.6 Summary of the Proposed Deep Learning Based Methods for Hyperspectral Image Classification

This chapter covers three Deep Learning (DL) based methods for Hyperspectral Image (HSI) classification, where spatial features are incorporated prior to classification.

Firstly, Shape-Adaptive Hyper-Segmentation based Stacked Auto-Encoder (SAHS-SAE) method is developed for HSI classification. The proposed SAHS-SAE is a two-step approach that integrates the effect of spatial features and replaces the fixed pixel-wise scanning window with an adaptive spatial structural window making it possible to fully exploit the capability of SAE. First, spatial contextual features are exploited via hyper-segmentation (HS) and are incorporated into multi-layer SAE for effective HSI classification. Unlike previous SAE-based feature extraction methods, SAHS-SAE approach develops a shape-adaptive neighborhood structural area for each sample pixel instead of using a square window of fixed size. Size and shape of the structural region can be flexibly adapted according to the actual HSI spatial structures, which results in an effective and adaptive exploitation of spatial contexts. SAHS-SAE then uses the multi-layer SAE to effectively exploit the HS-based spatial features and selected spectral features within and among segmented regions.

In the second proposed method, Deep Belief Network (DBN) is exploited for the adaptive window based spatial feature incorporation for HSI classification. In this method, first, spatial contextual features are extracted, like the last method, and then spatially similar region, along with the spectral features is fed into a DBN network to exploit the more effective spectral-spatial features for HSI classification.

In the third proposed HSI classification method, PCANet-based deep learning model is exploited with shape-adaptive spatial features for HSI classification. This approach involves a simplified deep learning SANet approach for HSI which studies the state-of-the-art deep learning model and its concerns. The exceptional learning power helps us to overlook the complex layered architecture and spatial information, which restricts the capability of deep learning in remote sensing situations. Therefore SANet is preferred; in this network-based architecture PCA filters take over the inner-layer convolution filter which eludes the optimizing process and reduces complexity. Furthermore, spatial information is overburdened by hyper-segmentation and its synthesizes with spectral characteristics. The multi-layer deep learning network is utilized by SANet, which uses the division-based spatial features effectively and also the selected spectral features present amid the segmented regions. These results performed on challenging training data depict better performance in structural regions for small areas. The further challenge is to improve performance by scaling down the training samples.

Extensive experimental analysis of AVIRIS and ROSIS sensors based on diverse HSI datasets demonstrate that proposed DL-based approaches produce better performance and outperform existing well-known and widely used HSI classification approaches, especially in images with small spatial structures. Furthermore, proposed methods, as one of the robust feature extractor, works well in heterogeneous regions, particularly for complex urban scenes.

References

1. Rajan S, Ghosh J, Crawford MM (2008) An active learning approach to hyperspectral data classification. IEEE Trans Geosci Remote Sens 46(4):1231–1242
2. Chen Y, Jiang H, Li C, Jia X, Ghamisi P (2016) Deep feature extraction and classification of hyperspectral images based on convolutional neural networks. IEEE Trans Geosci Remote Sens 54(10):6232–6251
3. Yuen PW, Richardson M (2010) An introduction to hyperspectral imaging and its application for security, surveillance and target acquisition. Imaging Sci J 58(5):241–253
4. Lacar F, Lewis M, Grierson I (2001) Use of hyperspectral imagery for mapping grape varieties in the Barossa Valley, South Australia. In: IEEE 2001 international geoscience and remote sensing symposium, 2001. IGARSS'01, vol 6. IEEE, pp 2875–2877
5. Van Der Meer F (2004) Analysis of spectral absorption features in hyperspectral imagery. Int J Appl Earth Obs Geoinf 5(1):55–68
6. Hege EK, O'Connell D, Johnson W, Basty S, Dereniak EL (2004) Hyperspectral imaging for astronomy and space surveillance. In: Optical science and technology, SPIE's 48th annual meeting. International Society for Optics and Photonics, pp 380–391
7. Malthus TJ, Mumby PJ (2003) Remote sensing of the coastal zone: an overview and priorities for future research
8. Bioucas-Dias JM, Plaza A, Camps-Valls G, Scheunders P, Nasrabadi N, Chanussot J (2013) Hyperspectral remote sensing data analysis and future challenges. IEEE Geosci Remote Sens Mag 1(2):6–36
9. Hughes G (1968) On the mean accuracy of statistical pattern recognizers. IEEE Trans Inf Theory 14(1):55–63
10. Landgrebe D (2002) Hyperspectral image data analysis. IEEE Signal Process Mag 19(1):17–28
11. Kang X, Li S, Benediktsson JA (2014) Spectral–spatial hyperspectral image classification with edge-preserving filtering. IEEE Trans Geosci Remote Sens 52(5):2666–2677
12. Pal M, Foody GM (2010) Feature selection for classification of hyperspectral data by SVM. IEEE Trans Geosci Remote Sens 48(5):2297–2307
13. Plaza A, Plaza J, Martin G (2009) Incorporation of spatial constraints into spectral mixture analysis of remotely sensed hyperspectral data. In: 2009 IEEE international workshop on machine learning for signal processing, 2009. MLSP 2009. IEEE, pp 1–6
14. Qian Y, Ye M (2013) Hyperspectral imagery restoration using nonlocal spectral-spatial structured sparse representation with noise estimation. IEEE J Sel Top Appl Earth Obs Remote Sens 6(2):499–515
15. Li J, Marpu PR, Plaza A, Bioucas-Dias JM, Benediktsson JA (2013) Generalized composite kernel framework for hyperspectral image classification. IEEE Trans Geosci Remote Sens 51(9):4816–4829
16. Zhou Y, Peng J, Chen CP (2015) Extreme learning machine with composite kernels for hyperspectral image classification. IEEE J Sel Top Appl Earth Obs Remote Sens 8(6):2351–2360
17. Zhang Y, Prasad S (2015) Locality preserving composite kernel feature extraction for multi-source geospatial image analysis. IEEE J Sel Top Appl Earth Obs Remote Sens 8(3):1385–1392
18. Zhang Q, Tian Y, Yang Y, Pan C (2015) Automatic spatial–spectral feature selection for hyperspectral image via discriminative sparse multimodal learning. IEEE Trans Geosci Remote Sens 53(1):261–279
19. Ghamisi P, Benediktsson JA, Sveinsson JR (2014) Automatic spectral–spatial classification framework based on attribute profiles and supervised feature extraction. IEEE Trans Geosci Remote Sens 52(9):5771–5782

20. Ghamisi P, Benediktsson JA, Cavallaro G, Plaza A (2014) Automatic framework for spectral–spatial classification based on supervised feature extraction and morphological attribute profiles. IEEE J Sel Top Appl Earth Obs Remote Sens 7(6):2147–2160
21. Li J, Zhang H, Zhang L (2014) Supervised segmentation of very high resolution images by the use of extended morphological attribute profiles and a sparse transform. IEEE Geosci Remote Sens Lett 11(8):1409–1413
22. Song B, Li J, Dalla Mura M, Li P, Plaza A, Bioucas-Dias JM, Benediktsson JA, Chanussot J (2014) Remotely sensed image classification using sparse representations of morphological attribute profiles. IEEE Trans Geosci Remote Sens 52(8):5122–5136
23. Fauvel M, Benediktsson JA, Chanussot J, Sveinsson JR (2008) Spectral and spatial classification of hyperspectral data using SVMs and morphological profiles. IEEE Trans Geosci Remote Sens 46(11):3804–3814
24. Wang G, Hoiem D, Forsyth D (2012) Learning image similarity from Flickr groups using fast kernel machines. IEEE Trans Pattern Anal Mach Intell 34(11):2177–2188
25. Hinton GE, Salakhutdinov RR (2006) Reducing the dimensionality of data with neural networks. Science 313(5786):504–507
26. Krizhevsky A, Sutskever I, Hinton GE (2012) ImageNet classification with deep convolutional neural networks. In: Advances in neural information processing systems, pp 1097–1105
27. Yu D, Deng L, Wang S (2009) Learning in the deep-structured conditional random fields. In: Proceedings of the NIPS workshop, pp 1–8
28. Mohamed AR, Dahl G, Hinton G (2009) Deep belief networks for phone recognition. In: NIPS workshop on deep learning for speech recognition and related applications, vol 1, p 39
29. Chen Y, Lin Z, Zhao X, Wang G, Gu Y (2014) Deep learning-based classification of hyperspectral data. IEEE J Sel Top Appl Earth Obs Remote Sens 7(6):2094–2107
30. Gómez-Chova L, Tuia D, Moser G, Camps-Valls G (2015) Multimodal classification of remote sensing images: a review and future directions. Proc IEEE 103(9):1560–1584
31. Tao C, Pan H, Li Y, Zou Z (2015) Unsupervised spectral–spatial feature learning with stacked sparse autoencoder for hyperspectral imagery classification. IEEE Geosci Remote Sens Lett 12(12):2438–2442
32. Chen Y, Zhao X, Jia X (2015) Spectral–spatial classification of hyperspectral data based on deep belief network. IEEE J Sel Top Appl Earth Obs Remote Sens 8(6):2381–2392
33. Mughees A, Chen X, Tao L (2016) Unsupervised hyperspectral image segmentation: merging spectral and spatial information in boundary adjustment. In: 2016 55th annual conference of the society of instrument and control engineers of Japan (SICE). IEEE, pp 1466–1471
34. Mughees A, Chen X, Du R, Tao L (2016) Ab3c: adaptive boundary-based band-categorization of hyperspectral images. J Appl Remote Sens 10(4):046009
35. Li W, Du Q (2014) Gabor-filtering-based nearest regularized subspace for hyperspectral image classification. IEEE J Sel Top Appl Earth Obs Remote Sens 7(4):1012–1022
36. Tan K, Li E, Du Q, Du P (2014) Hyperspectral image classification using band selection and morphological profiles. IEEE J Sel Top Appl Earth Obs Remote Sens 7(1):40–48
37. Li J, Bioucas-Dias JM, Plaza A (2011) Hyperspectral image segmentation using a new Bayesian approach with active learning. IEEE Trans Geosci Remote Sens 49(10):3947–3960
38. Chen Y, Nasrabadi NM, Tran TD (2013) Hyperspectral image classification via kernel sparse representation. IEEE Trans Geosci Remote Sens 51(1):217–231
39. Li J, Bioucas-Dias JM, Plaza A (2013) Spectral–spatial classification of hyperspectral data using loopy belief propagation and active learning. IEEE Trans Geosci Remote Sens 51(2):844–856
40. Hu W, Huang Y, Wei L, Zhang F, Li H (2015) Deep convolutional neural networks for hyperspectral image classification. J Sens 2015
41. Melgani F, Bruzzone L (2004) Classification of hyperspectral remote sensing images with support vector machines. IEEE Trans Geosci Remote Sens 42(8):1778–1790
42. Ji R, Gao Y, Hong R, Liu Q, Tao D, Li X (2014) Spectral-spatial constraint hyperspectral image classification. IEEE Trans Geosci Remote Sens 52(3):1811–1824

43. Hinton GE (2002) Training products of experts by minimizing contrastive divergence. Neural Comput 14(8):1771–1800
44. Chan T-H, Jia K, Gao S, Lu J, Zeng Z, Ma Y (2015) PCANet: a simple deep learning baseline for image classification? IEEE Trans Image Process 24(12):5017–5032
45. Benediktsson JA, Ghamisi P (2015) Spectral-spatial classification of hyperspectral remote sensing images. Artech House, Boston
46. Mou L, Ghamisi P, Zhu XX (2017) Deep recurrent neural networks for hyperspectral image classification. IEEE Trans Geosci Remote Sens

Chapter 6
Multi-Deep Net Based Hyperspectral Image Classification

This chapter presents Deep Learning-based hyperspectral image (HSI) classification techniques where the complexity of HSI is addressed in a unique way. This is phase 4 and phase 5 in our developed framework as shown in Fig. 6.1. A complete description of each phase is depicted in Chap. 1, Fig. 1.3.

The HSI analysis can be divided into two parts. In the first part, spectral segmentation is proposed to proficiently decrease and distribute the complication, and effusively explore the presented spectral and spatial data for enhanced feature mining by keeping the scarcity of fewer training data that are accessible in the hyperspectral scenario. This chapter presents the segmented DBN approach. Here, local deep belief networks are employed for every division of the spectral bands. This procedure comprises twofold major phases. Firstly, flexible edge-regulation rooted segmentation is executed to explore rich spatial resolution present in the hyperspectral image. Secondly, spectral division is executed, where spectrally analogous adjacent channels are convened together and a deep belief network is executed to each divided band cluster distinctly. Locally employing DBN rooted feature mining to every individual cluster of channels decreases the calculation intricacy and simultaneously results in better features and hence improved classification accuracy is obtained as compared to the existing HSI classification techniques.

In the second part of this chapter, methods based on the incorporation of spatial information after the classification are developed. Two main DL architectures SAE and DBN are used for this purpose. A detailed analysis along with the experimental results and comparisons is presented in the second part of the chapter. The subject classification approaches utilize spatial information once pixel-wise classification is applied with the assistance of a resolution instruction. In such techniques, contextual data is exploited with the help of numerous methods, for example, morphological information [1], Attribute Profiles [2], segmentation [3, 4], and Markov random fields (MRFs). The final explored contextual data and pixel-wise classification rooted infor-

L. Tao and A. Mughees, *Deep Learning for Hyperspectral Image Analysis and Classification*, Engineering Applications of Computational Methods 5,
https://doi.org/10.1007/978-981-33-4420-4_6

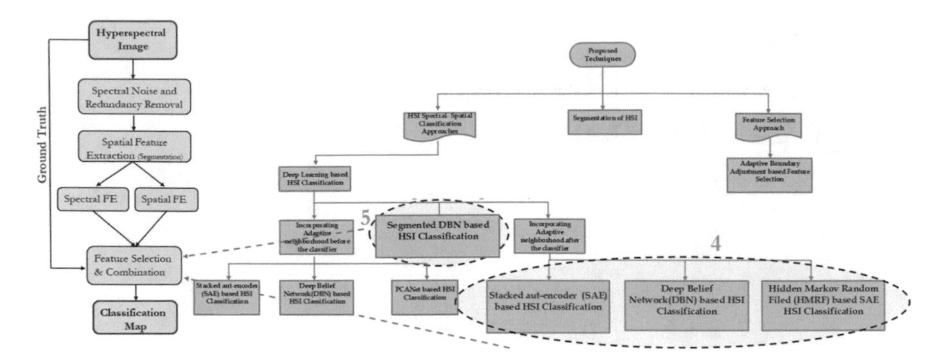

Fig. 6.1 Fourth phase aligned with the framework toward the classification of hyperspectral remote sensing images

mation is integrated collectively to deliver the ultimate classification performance. In the aforementioned approaches, particular class assignment to individual pixels is determined on the basis of pixel-wise classification performance for the target pixel, and neighboring contextual information attained with the help of segmentation. The precise region detection of the hyperspectral image marks this technique as one of the improved methods in terms of performance. The developed technique initially explores the DL architectures for mining feature presentations of the hyperspectral scene. Subsequently, a similar cell segmentation procedure is deployed through the HMRF-EM rooted approach [5] and through the proposed segmentation technique. Lastly, the majority vote (MV) technique is engaged to incorporate the outputs of pixel-based spectral classification and grouping of pixels based on the spatially similar regions which results in the accurate classification of HSI. The performance of this approach is substantiated through the accurate results and its comparison to the existing HSI classification techniques.

6.1 Multi-Deep Belief Network-Based Spectral–Spatial Classification of Hyperspectral Image

As depicted in previous chapters advancements in the optical innovation has empowered hyperspectral scene (HSI) acquiring sensors to procure the reflectance in many distinctive wavelengths from the earth's surface going from visible to the infrared range as several continuous narrow spectral bands. Each band in HSI is a 2D spatial guide with varying spectral signature. Thus, every pixel in HSI is a vector of the estimation of spectral attributes of a specific spatial situation in the scene [6]. Such a wide scope of otherworldly data is incredibly valuable in numerous applications, for example, mineral location [7], exactness cultivating [8], urban arranging, ecological observing [9], and target recognition and observation [10]. Be that as it may, such an abundance of data in HSI accompanies more difficulties, cost of multifaceted nature, and unwanted factual and geometrical properties [11] which requires advanced methods [12]. High dimensionality in spectral space [13] combined with the curse of dimensionality [14] and constrained training examples makes HSI analysis a difficult issue [14].

Advancements in the optical innovation has empowered hyperspectral scene (HSI) acquiring sensors to procure the reflectance in many distinctive wavelengths from the earth's surface going from visible to the infrared range as several continuous narrow spectral bands. Each band in HSI is a 2D spatial guide with varying spectral signature. Thus, every pixel in HSI is a vector of the estimation of spectral attributes of a specific spatial situation in the scene [29]. Such a wide scope of otherworldly data is incredibly valuable in numerous applications, for example, mineral location [28], exactness cultivating [21], urban arranging, ecological observing [13], and target recognition and observation [32]. Be that as it may, such an abundance of data in HSI accompanies more difficulties, cost of multifaceted nature, and unwanted factual and geometrical properties [20] which requires advanced methods [16]. High dimensionality in spectral space [6] combined with the curse of dimensionality [14] and constrained training examples makes HSI analysis a difficult issue [2].

High dimensionality of hyperspectral information is a significant issue. As taking care of numerous spectral channels simultaneously prompts intricacy, faults and keeps DBN away from the extraction of progressively theoretical and profound rich features with respect to every pixel in HSI, there exist more than 200 spectral bands and in this way an equivalent number of units are executed at the information layer to deal with this rich information. Moreover, parameters involved in the inner layers become more complex because of the number of input units, which makes the whole system more complex. In addition, increment in the spectral band and rich spectral information likewise requires the expansion in inspected preparation of labeled training information at a similar rate. In any case, the accessibility of training data is a significant bottleneck in HSI. The awkwardness between the high-dimensional information and test data prompts the Hughes phenomenon [15]. Generally, effective and robust HSI analysis execution relies upon the effective use of high-dimensional information by keeping the processing within the limits of constrained available test information as each band in the hyperspectral 3D square exhibits important data about each class in HSI. There are two principle measurement decrease approaches used to deal with the high dimensionality, specific feature determination, i.e., feature selection (FS) and feature extraction (FE). Feature determination also called as the band choice strategy chooses a few bands out of a hypercube to represent the entire HSI 3D cube [8, 26]. On the other hand, the FE approach accomplishes the dimensional decrease by anticipating the novel data to a lower dimensional element space, for example, with the help of techniques like principal component analysis [30], autonomous segment analysis [3], and most extreme noise division [33]. The FE procedure changes the information into feature space and looks for the subset of viable and suitable features in the feature space [18, 27]. Nevertheless, the significant downside of dimensional decrease systems is the loss of significant, important data as each spectral channel among several channels contains important data about every material. Thus, dimensionality decrease procedures bring about a trade-off in the accuracy of hyperspectral image classification.

Lately, deep learning designs have exhibited their ability in HSI analysis for successful feature extraction. These models have demonstrated to deliver empowering execution in scene, sound, and language processing [15–18]. As mentioned in the previous chapter, deep learning architectures are the latest advancement in the neural system with numerous handling layers that are fit for extricating progressively unique and profound invariant features which results in improved classification performance [4, 19–23]. Of late, some DL models have been actualized on HSI. Profound learning-based stacked auto-encoder (SAE) is proposed in [13] and deep belief network (DBN) is proposed in [24]. These techniques demonstrated to acquire profound features from the HSI scene. Of late, there is an expanding interest for incorporating the spatial contextual information alongside the spectral data as it has of late been shown that joining spectral and spatial information altogether improves the performance results [4, 19–23]. In [1], the reconciliation of spatial data through morphological profiles and unique spectral data is performed. Loopy belief propagation is used to explore the spectral and spatial characteristics in [25]. Kernel-based scarce depiction is utilized for HSI in [26].

Numerous classifiers have been researched in HSI analysis. Conventional HSI classification strategies are famous for considering the spectral properties only, for example, logistic regression [27], Bayesian classifier, random forests, neural systems, K-nearest neighbor classifier (K-NN), conditional random fields (CRFs), inadequate coding, and support vector machines (SVM). SVM performed well due to its capacity to deal with the high-dimensional information. Nevertheless, these methods represent a few confinements. The full association between the diverse concealed layers makes the profound learning-based HSI characteristics extraction process extremely difficult and time consuming as HSI involves many spectral channels, and numerous neurons are required to assess the information and approximate parametric values from all the accessible channels simultaneously in a similar initiation work. This network of intricate associations results in the absence of appropriate abstraction. Additionally, layers' full associations require more testing labeled information to prepare the parameters, which is a significant restriction in HSI as constrained labeled data is accessible for HSI. SAE and DBN can't extricate the spatial data effectively as a fixed estimated window is utilized to separate the spatial data which may contain various classes or a subset of an equivalent class.

In this segment of the chapter, a spectrally adapted divided deep belief network (SAS-DBN) is presented which performs spectral division to proficiently decrease and split the complex nature of data effectively and completely explore the accessible spectral and spatial data for better characteristics extraction by keeping the algorithm within the limitation of restricted availability of the training data. This part of the chapter presents the fragmented DBN system where nearby DBNs are applied to each portion of the available bands. It comprises two principal steps: first, versatile limit alteration-based segmentation is performed to explore the spatial data; furthermore, spectral division is performed where bands which are similar in nature from a spectral point of view are assembled and afterward, DBN is applied on each spatial–spectral fragmented channel groups independently. Indigenously employing DBN rooted characteristic mining to every set of channels decreases the calculation intricacy and concurrently consequences into improved characteristics and therefore accurate classification results are acquired. The major impact of this research can be itemized as follows:

- Spectral data in every band is completely explored. Spectral comparable adjacent channels are assembled and DBN is employed to resulting fragments of the range for proficient feature extraction that partitions the multifaceted nature and furthermore permits nearby extraction of a spectral characteristic within the limitation of fewer training data. Henceforth, improved characterization is acquired.
- Spatial data is completely explored by swapping the fixed size spatial window with hyper-segmented regions which flexibly varies the scope and formation as indicated by the original structure in HSI.

The primary focal point of this work is to explore ways by which to productively use all the significant data given in the spectrum without trading off the computational intricacy for HSI and without decreasing the bands of the hyperspectral information.

6.1.1 Spectral-Adaptive Segmented DBN for HSI Classification

The developed technique explores the spectral characteristics with the help of a flexible spectral channel grouping rooted deep belief network to enrich and make the feature extraction procedure more accurate, where HSI channels that are greatly interrelated and associated are assembled collectively and are given as an input to one of the many DBN architectures to excerpt additional deep characteristics of data. A complete flowchart of the developed methodology is depicted in Fig. 6.2. The characteristics excerption procedure should consider the accompanying two spectral–spatial components [13]:

- There is a high likelihood that HSI pixels with indistinguishable spectral marks add to a similar class name.
- Neighboring pixels in the spatial area that are exceptionally important in spatial marks ought to likewise add to a similar class name.

Above mentioned two aspects clearly explain that the possibility of pixels which spatially and spectrally possess a great association and are interrelated and their probabilty of belonging to the same class is very high. In order to efficiently apply the mentioned aspects, DBN is advanced to explore spectral and contextual characteristics by switching the orthodox rigid-sized contextual window with a flexible region-based window. Additionally, grouping bands which are similar in spectral characteristics for DBN increase the accuracy performance. The complete picture of the developed method is depicted in Fig. 6.2. In the first stage, a contextually flexible region-based approach is employed [28] to partition the HSI scene in contextually

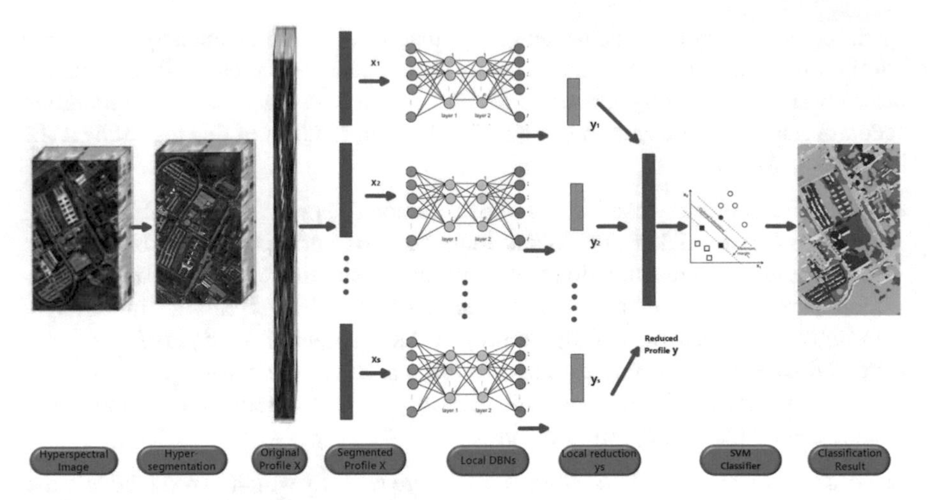

Fig. 6.2 Framework of the SAS-DBN process

same areas. Subsequently, bands which are spectrally identical in nature are assembled into different spectrally associated groups and each group is then given as an input to each of the local DBNs for feature mining.

6.1.1.1 Hyper-segmentation-Based Spatial Feature Extraction

As deliberated above and in the previous chapter, taking spatial information into account may end up in an accurate classification outcome. Past studies [29] show that the researchers have explored the contextual data for hyperspectral image classification by utilizing a standing window shape. Conversely as discussed in the previous chapter, applying the standing window shape confines the mining of contextual characteristics, which leads to reducing the accuracy afterward. Since a particular-sized window could comprise more than an object, there is also a possibility that it consists of pixels that are only the subsection of the object in the HSI scene as depicted in Fig. 5.2. In the flexible boundary rooted region-segmentation technique which was designed in Chap. 4, adjacent spatial data that contain contextually comparable features are assembled collectively. The flexible edge detection rooted technique designed in Chap. 4 is applied to effectively partition the hyperspectral image scene into contextually comparable areas. Therefore, contextual stability is guaranteed by utilizing the abovementioned region-based boundary evolution technique [28] which partitions the hyperspectral scene by utilizing confined contextual consistency, wherever the extent and outline of an object is flexibly attuned, built on the real edges. The following are the major phases for segmentation:

1. Split the hyperspectral scene into m starting hexagonal fragments.
2. Loop over the adjacent pixels till no pixel is assigned a new segment.

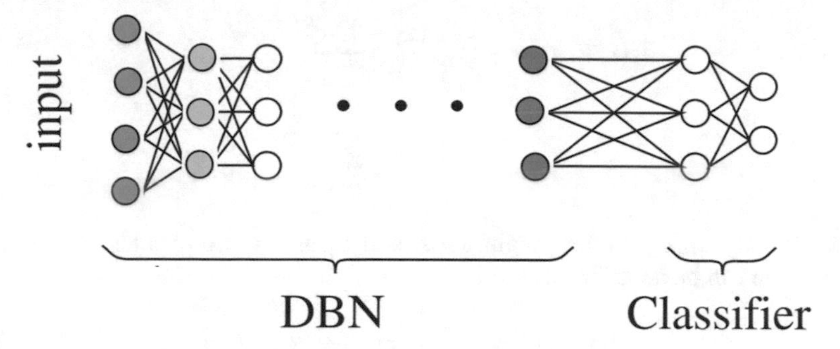

Fig. 6.3 Framework of the DBN-based pixel-wise classification

- Calculate the major class of each initial segment.
- Calculate the gradient of current boundary pixels of each segment.
- Calculate the straightness factor for each segment boundary.
- Based on the equation, compute the degree of resemblance, between each class centroid and its boundary.
- Advance the edge based on the resemblance degree.

3. For each channel gathering, dole out a load to every limit pixel as indicated by specific criteria to grant pixels that mirror a real limit.

The itemized methodology for the hyper-division is outlined in the following figure.

6.1.2 Spectral–Spatial Feature Extraction by Segmented DBN

6.1.2.1 DBN

A profound conviction net is a development of generative neural system rooted learning components every one of which is a Restricted Boltzmann Machine that comprises an information layer to get information and a concealed layer that figures out how to recognize spectral characteristics that can retrieve higher-abstract connections in the information as shown in Fig. 6.3. The double RBM layers are associated with a framework of evenly weighted associations, w, with no interlaced associates. This limitation makes the concealed units restrictively free. A joined development of a function for hidden neuron u and input neuron x is provided by [24]:

$$E(x, u, \theta) = -\sum_{j=1}^{n} \frac{(x_j - b_j)^2}{2\sigma^2} - \sum_{i=1}^{m} a_i u_i \\ - \sum_{j=1}^{n} \sum_{i=1}^{m} w_{ji} \frac{x_j}{\sigma_i} u_i \tag{6.1}$$

where $\theta = b_j, a_i, w_{ji}$ is the weight units, and b_j and a_i are bias terms. The joint distribution can be described as

$$P(x, u, \theta) = \frac{a}{N(\theta)} exp(-E(x, u, \theta)) \tag{6.2}$$

Here, $N(\theta)$ is the normalization parameter. The framework provides a probabilistic parameter to every input vector with $E(x, u, \theta)$. The conditional distribution is provided by

$$P(u_i|x; \theta) = g\left(\sum_{j=1}^{n} w_{ji} x_j + a_i\right) \tag{6.3}$$

$$P(x_i|u; \theta) = N\left(\sum_{i=1}^{m} w_{ji} u_i \sigma_j^2 + b_j\right) \tag{6.4}$$

where σ is predictable error, and $N(.)$ is the Gaussian distribution. The parameters are erudite by contrastive divergence (CD) [30] and are modified by

$$\Delta w_{ij} = \psi(x_i u_{j_{data}} - x_i u_{j_{reconstruction}}) \tag{6.5}$$

where ψ is the learning rate, and x_i and u_j are the initial input units and the inner layers' neurons, correspondingly .

The power of RBM lies is in the reconstruction of the real data. In the restoration stage, the useful info learned by the inner neuron through training is exploited. The erudite values are deliberated proficiently if this procedure can recover the unique real information. This shows that inner neurons have saved sufficient information about the real given data.

Only one layer is not appropriate to explore the useful information present in the provided hyperspectral channels. Thus, the output of the first RBM layer after training can be exploited by giving it to the second layer as this input is a more refined and learned set of neurons that can play a more important role and can help the system for more deep feature mining. Hence, multiple layers are integrated over one another and can be given training with labeled samples to formulate a deep belief network. The complete procedure can be briefed as follows:

1. Available training data for a specific hyperspectral scene is utilized for training the singular layer of the deep belief network. Contrastive divergence (CD) [30] is exploited for training the RBM.
2. The result of the first RBM layer is provided as an input to the succeeding layer, which independently goes under the process of training.
3. Loop through steps 1 and 2 for as many repetitions as required for the number of layers.
4. Adjust all the learned values and try to get the finest possible values in the presence of the fewer available sampled data.
5. The developed framework utilized SVM as an algorithm to classify the final features obtained from the final layer of RBM.

In summarized form, initially, hyper-region-based segmentation is performed to obtain the contextual comparable region. These regions along with the spectral band data are provided as input to the deep belief network. At that point, DBN is applied to get the unique characteristics of the input information through multi-layer DBN. Lastly, LR is used to characterize and mark the pixels dependent on learning features as depicted in Fig. 6.4. This process is done for every grouped spectral band as depicted in Fig. 6.4.

6.1.2.2 Segmented DBN

In conventional utilization of DBN in HSI, all the band groups are dealt with in the same way and also simultaneously. This results in drastic escalation in the amount of learned parameters and also increases complications. The major reason behind this enhanced complication is that the neurons of the first layers are attacked with thousands of raw input spectral values, and this becomes more and more complex in the following layers. In addition to this, the association and relationship among different spectral channels are not incorporated and channel spectral values are given blindly to these deep learning-based architectures. Consequently, looking into these loopholes in the developed technique addresses this area by employing a deep belief network to the individual assembled associated spectral bands. This serves two purposes: firstly, the association and connection that exist between bands are incorporated as only those bands are assembled which have high associativity and relationship among each other. Secondly, complexity and intricacy are reduced manifold as now only the subsection of the total bands is given as an input to the local DBN. This conceptual idea of employing a specific deep learning architecture to the diverse sections is already tried with good performance in various feature mining techniques including [31] and also with various adaptions in [32].

This developed concept is explained comprehensively and thoroughly with the help of images in Fig. 6.2. Here, the provided amount of bands $\mathbf{x}$ for a particular hyperspectral scene are partitioned into $\mathbf{S}$ various groups $\mathbf{x}_s, s \in [1, S]$ and afterwards deep belief network is autonomously employed to every partitioned region of bands.

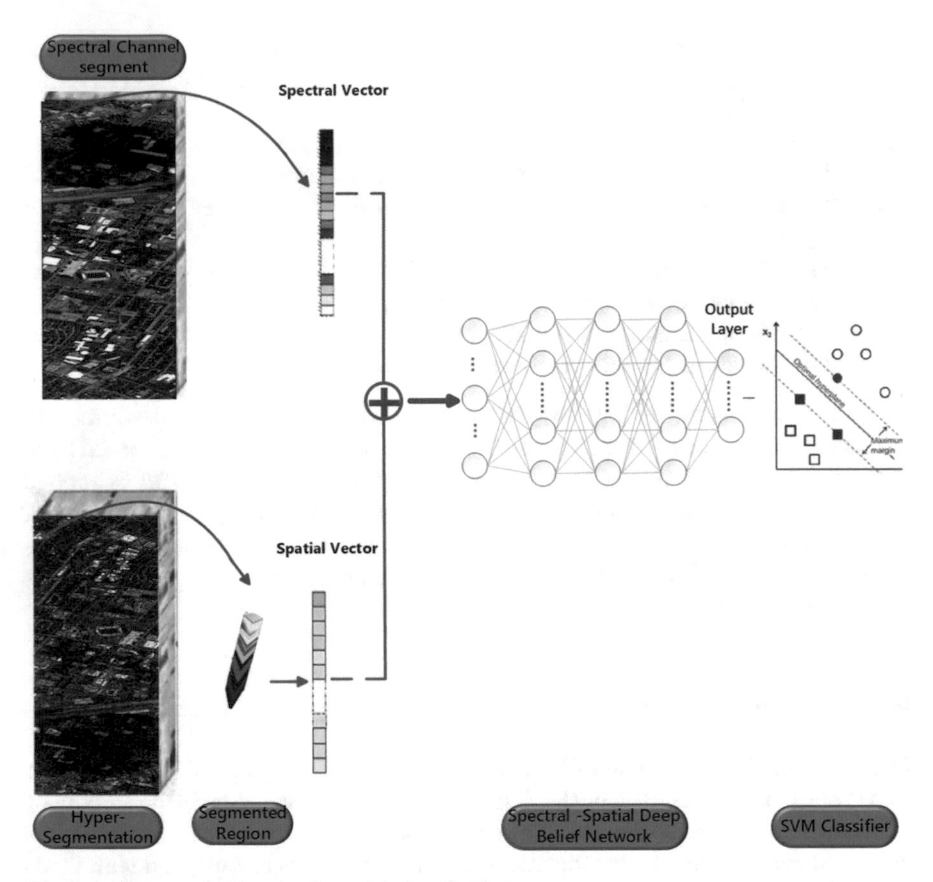

Fig. 6.4 Framework of spectral–spatial classification

Since every individual DBN is employed to a very small assembled number of bands, this leads to a need of very few number of hidden neurons which in turn results in the formation of fewer number of parameters to learn and hence less training data is required. This whole process results in the simpler employability of DBN in comparison to the DBN employed in the traditional way. Furthermore, due to less bands for each individual DBN, variant and more effective characteristics from each group of bands can be attained. Lastly, distinguishable characteristics obtained from each DBN architecture y_s are combined ($\sum_{s=1}^{S} = F_s = F$) to formulate the final set of the feature vector.

In order to evaluate the degree of associativity, i.e., the extent of similarity in terms of spectral signatures of a particular hyperspectral scene, a correlation matrix between channels is calculated. This matrix proves to be very effective for evaluating the degree of connectivity between different bands. Channels with a maximum degree of similarity depicted through correlation matrix are assembled [32] to form one group of similar bands. In this way, different groups are formed for SAS-DBN. This correlation matrix is evaluated through a covariance matrix. Mathematically, this can be demonstrated as

$$Cov = F\{(p - E(p))(p - F(p))^T\} \tag{6.6}$$

Here, $F\{.\}$ is the expectancy function. Rooted from this function, every value in the associativity matrix can be calculated as

$$Corr(a, b) = \frac{Cov(a, b)}{\sqrt{Cov(a, a)Cov(b, b)}} \tag{6.7}$$

Here, $Cov(a, a)$ and $Cov(b, b)$ present the association of the ath and ath bands, correspondingly. Consequently, $Corr(a, b)$ presents the association between the ath and bth spectral bands in the hyperspectral image dataset. Henceforth, the entire associative matrix depicts an association between the spectral bands in the hyperspectral scene dataset. Hence that proves that associativity relation can be effectually deployed to formulate the groups of similar bands.

6.1.3 Experimental Results and Performance Comparisons

To measure and access the performance of designed method, various experiments were conducted on available HSI datasets including AVIRIS sensor based natural images, Houston University and Pavia University hyperspectral scene datasets. A comprehensive explanation of every utilized HSI scene is described in Chap. 2, Sect. 2.3.

6.1.4 Experimental Setup

The analyses of the developed methodology were led on a 4.0 GHz processor with NVIDIA GeForce GTX 970 and on a Windows 7 working framework. For coding implementation, the Theano framework was employed. Traditional DBN can be utilized from multiple points of view in terms of the number of hidden units and the number of layers. Customized DBN gives best execution performance with 2–6 layers and 20–6o hidden units as recommended in [24]. The hidden layer neurons are equivalent to the quantity of characteristics that are necessary. From these outcomes, we utilized a 42-layer DBN comprising 40 inner neurons each. Overall, 3 layers for Pavia college and Houston college datasets were exploited consisting of 1 input layer and 2 inner layers with the arrangement of $x40... \smile F$. Here, x relates to the number of bands as a contribution to the DBN, and F is the quantity of characteristics required at the last layer. Increasing the number of hidden neurons or the number of layers doesn't necessarily have a positive effect on the performance and accuracy as experimented by [24].

6.1.4.1 Configuration for Segmented DBN

The proposed technique initially partitions the spectral band into several distinct groups to apply the DBN on each individual grouped set of bands independently. The association between spectral channels can be effectually calculated by utilizing an associativity matrix (correlation matrix (CM)) [32]. This association matrix is calculated through a covariance matrix. Any specific value in the association matrix presents the level of association among the particular bands. Let's take an example: suppose CM(n,m) signifies the equivalent relationship between nth and mth channels of a specific hyperspectral image. The complete association matrix carries the degree of similarity among the entire pair of channels in the HSI scene that can be proficiently explored to develop spectral channel groups which are also presented in Fig. 6.5. As endorsed by [32], the partitioned set of bands can be obtained from the major association collections resulting from the association matrix. Subsequently, Table 6.1 depicts the derived collection of spectral bands. The entire number of characteristics are also shared (5, 10, 15, 20) consistently between the groups.

6.1.5 Spectral–Spatial HSI Classification

Experimental outcomes of the designed methodology were given a performance comparison with the well-known state-of-the-art established and newly designed approaches, the support vector machine (SVM) [33], the newly established recurrent

Table 6.1 Segmented-DBN Configuration for Pavia University and Houston University datasets

Dataset	Grouped channels	Grouped channels	Hidden-layer units	Reduced features			
Pavia University	Segment S_1	1 – 22	$L_1 = 13$ $L_2 = F$	1	3	3	6
	Segment S_2	23 – 60	$L_1 = 13$ $L_2 = F$	2	3	6	7
	Segment S_3	61 – 103	$L_1 = 14$ $L_2 = F$	2	4	6	7
				5	10	15	20
Houston University	Segment S_1	1 – 76	$L_1 = 13$ $L_2 = F$	2	4	7	6
	Segment S_2	77 – 110	$L_1 = 14$ $L_2 = F$	1	3	7	6
	Segment S_3	111 – 144	$L_1 = 13$ $L_2 = F$	2	3	6	3
				5	10	15	20

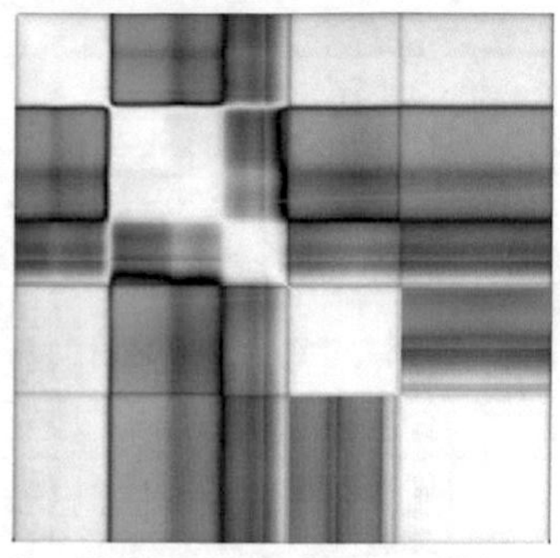
(a) Correlation matrix (white=1 black=0)

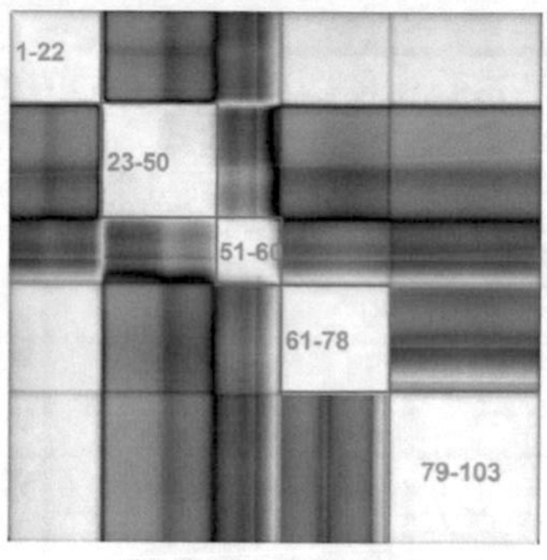

(b) Selected regions

Fig. 6.5 Correlation matrix and corresponding band grouping for the Pavia University dataset

neural network (RNN) [34], the deep belief network with LR (DBN-LR) [24], and lately designed convolutional neural network (CNN) [35].

The Hyperspectral Image scene comprising Houston University is believed to be one of the difficult and complex scenes because of containing contextual areas that are really minor in terms of the pixel size. In the distribution of samples related to testing and training, simply 10% of data arbitrarily selected from every class were utilized for training data while 50 characteristics were exploited in every scene for classification. Performance of the designed methods on each class of Houston University and Pavia University scenes is presented in Tables 6.2 and 6.3.

It can be clearly seen in the tables that the proposed SAS-based DBN supplied improved classification results in *Overall Accuracy*, *Average Accuracy*, and *Kappa Coefficient* in comparison to the mentioned 4 famous established approaches. While it can also be observed that the DBN-LR technique provided comparatively good results in classes 1 and 2, the possible cause could be the presence of very fewer training data for subject cases. The designed approach gave comparatively efficient results especially for classes that contained very small contextual objects. Overall, the classification results of the technique showed much better and improved performance in comparison to the compared techniques.

Figures 6.6 and 6.7 clearly present the visual performance with the aid of the resulting maps of the several listed compared approaches for the two scenes already mentioned. It can be clearly observed from the visual inspection and class-wise performance of the developed method that it achieved much enhanced and improved results in terms of individual and overall classification accuracy by intelligently integrating the contextual information firstly and also taking advantage of the grouped-level spectral information and applying DBN locally that really helped to extract the meaningful characteristics without putting too much burden on parameter learning and making it less complicated. Furthermore, this method also performed better at the places where other techniques showed scattered results.

Table 6.2 Classification accuracy (%) of each class for the Houston University dataset obtained by SVM [33], RNN [34], CNN [35], DBN-LR [24], and proposed SAS-DBN

Class	Training	Test	SVM	RNN	CNN	DBN-LR	SAS-DBN
1	125	1126	97.47	82.53	81.20	**99.20**	98.80
2	125	1129	98.32	83.36	83.55	**99.60**	99.0
3	70	627	99.37	**100.0**	99.41	**100.0**	**100.0**
4	124	1120	98.01	90.53	91.57	**99.60**	**99.60**
5	124	1118	96.01	97.82	94.79	**99.60**	**99.60**
6	33	292	**99.83**	93.01	95.10	97.2	98.81
7	127	1141	91.23	75.37	63.53	97.0	**98.11**
8	124	1120	86.23	42.36	42.64	97.8	**98.0**
9	125	1127	86.99	77.62	58.17	94.0	**95.11**
10	123	1104	91.42	57.63	41.80	97.4	**97.75**
11	124	1111	91.67	77.42	75.71	97.3	**97.95**
12	123	1110	87.05	69.74	84.15	95.2	**96.05**
13	47	422	78.16	66.32	40.00	88.0	**90.55**
14	43	385	97.42	**100.0**	98.79	**100.0**	**100.0**
15	66	594	99.49	95.98	97.89	**100.0**	**100.0**
Overall Accuracy			93.06	89.85	85.42	97.70	**98.35**
Average Accuracy			93.25	80.65	76.55	97.50	**98.06**
Kappa Coefficient			0.925	0.7606	0.7200	0.975	**0.9805**

Table 6.3 Classification accuracy of each class for the Pavia University dataset obtained by SVM [33], RNN [34], CNN [35], DBN-LR [24], and proposed SAS-DBN

Class	Training	Test	SVM	RNN	CNN	DBN-LR	SAS-DBN
1	597	6034	**97.50**	84.45	87.34	87.37	89.11
2	1681	16971	**97.70**	85.24	94.63	92.10	93.55
3	189	1910	78.53	54.31	**86.47**	85.57	**87.50**
4	276	2788	89.29	95.17	96.29	95.11	**97.35**
5	121	1224	98.77	**99.93**	99.65	99.74	99.19
6	453	4576	83.04	80.99	93.23	91.94	**93.85**
7	120	1210	64.58	88.35	93.19	92.21	**93.55**
8	331	3351	86.90	**88.62**	86.42	87.02	88.05
9	85	862	99.92	99.89	**100.0**	**100.0**	**100.0**
Overall Accuracy			92.04	88.85	92.56	91.18	**93.15**
Average Accuracy			88.47	86.33	93.02	92.34	**93.06**
Kappa Coefficient			0.903	0.8048	0.9006	0.8828	**0.9105**

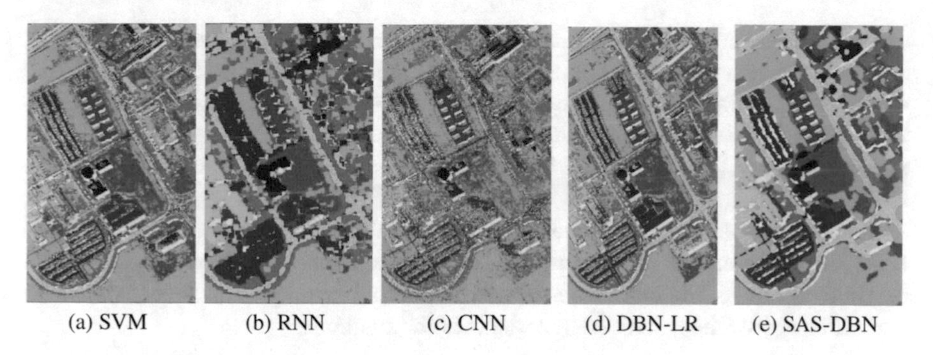

Fig. 6.6 Classification maps of various techniques for Pavia University

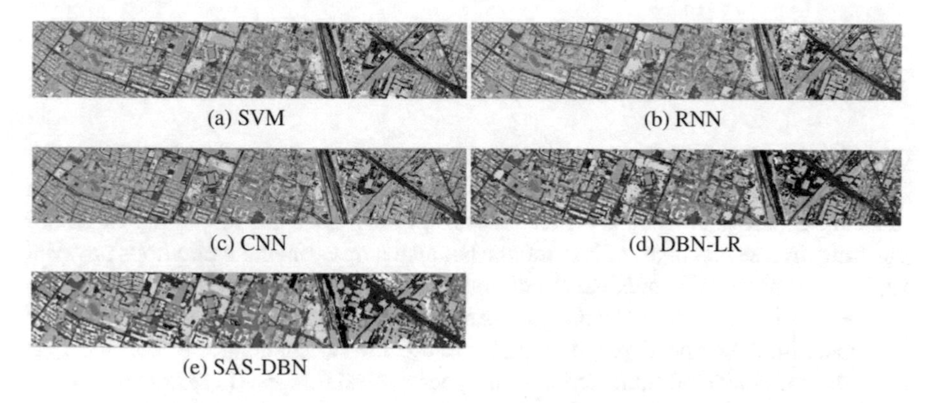

Fig. 6.7 Classification maps of various techniques for Houston University

6.1.6 Summary of the Proposed Multi-Deep Net-Based Hyperspectral Image Classification Method

In this part of the chapter, a well-researched and designed approach for hyperspectral image classification was presented which has addressed two main drawbacks of existing techniques. Firstly, it addressed the integration of contextual information by replacing the stationary-sized window with a flexible window that is based on the actual object boundary in the scene. Secondly, it addresses the complexity issue with so many spectral channels that go as an input to the deep learning architecture that in turn increases the complexity manifold and also results in hundreds and thousands of more learnable parameters; this issue was resolved with the help of grouping of the bands with similar spectral characteristics hence taking into account the spectral similarity as well and then using the local DBN-based deep learning architecture for feature mining. These two factors not only improve the performance of classification accuracy tremendously but also decrease the number of parameters hence reducing the complexity. In this approach, contextual information is gathered through assembling the contextually associated pixels by employing the flexible edge-based segmentation method that was already developed in previous chapters. Then spectral channels are also grouped together by measuring the degree of associativity among them by exploiting the associativity matrix. These characteristics are fed into the local DBN for feature mining and the resulting features from all the spectral groups are integrated at the end and fed into the classifier to label each pixel based on the feature.

A comprehensive experimental setup was established to demonstrate the capability of the developed framework and to compare the efficiency with well-known existing methods. The outcome of the experimental setup proves the effectiveness and efficiency of the technique over the existing methods.

6.2 Hyperspectral Image Classification Based on Deep Auto-Encoder and Hidden Markov Random Field

Current development in hyperspectral devices allows acquiring enormous diversity of data in hundreds of spectral bands. The enlarged quantity of spectral bands in the presence of adequate contextual resolution not only escalates the ability of distinguishing diverse resources of importance but also empowers the meticulous physical exploration of trivial contextual structured objects [19] in a given image.

The effective hyperspectral image classification (HSI) finds its interest in scientific and operational perspective. However, there are several challenges to consider: (1) Multifaceted statistical characteristics of hyperspectral images, (2) restricted amount of training samples data, and (3) imbalance between given labeled data and number of bands. Several hyperspectral classification approaches have been designed in the past many years due to their significance. In the initial period of the HSI scene classification, several approaches established on individual pixel rooted machine learning concepts were designed. These methods try to classify every pixel value of the specified scene independently without incorporating the weightage of contextual neighboring association of pixel which plays an imperative role. Traditional pixels based methods include logistic regression (LR), K-nearest neighbors (KNN), and maximum likelihood [24]. Secondly, characteristic mining [36] and characteristic selection [37] techniques were also suggested to handle the Hughes phenomenon. This includes a classifier that utilizes the compact characteristics to perform classification rather than processing the actual dimensional data.

Support vector machine (SVM) is considered as a good classifier for HSI classification, as it is less sensitive to the high dimensional of the data and produces good results even on less training data. However, a big limitation of the SVM is its incapability to consider spatial information for classification and classify just based on the spectral information. In recent times, deep learning-based classifiers [38] are introduced for HSI classification

After realizing the impact of contextual information along with spectral data, for HSI scene analyses, recent developed techniques based on spectral and spatial incorporation of features have proved to be very effective and has demonstrated major advancement in analyses [28, 39, 40]. These methods are based on different techniques to integrate the contextual and spectral characteristics of the hyperspectral scene [4, 23, 41]. These classifying approaches utilize contextual characteristics after pixel-based classification with the help of a decision rule. Spatial-contextual features can be mined with the aid of numerous approaches, such as morphological processing [1], Attribute Profiles [2], segmentation [3, 4], and Markov random fields (MRFs). The final mined contextual information and pixel-based classification outcome is integrated to deliver the concluding classification outcomes. MRF architectures are believed to be expectation-based architectures [42] that can be demarcated as a 2D stochastic procedure over distinct pixel frames [42]. Hidden Markov random field (HMRF) have shown to be an effectual and operative procedure for merging contextual characteristics into the classification [43].

Typical classifiers such as SVM and LR comprise distinct layers, though SVM consisting of kernels can have at most dual layers [44]. In current times, multiple-layer deep learning rooted hyperspectral scene classification has established a substantial improvement [13, 24]. To mine the shallow characteristics of a hyperspectral scene including the contextual characteristics to enhance the performance, it is vital to exploit a combination of numerous deep learning rooted models with the spatial characteristic rooted approaches like segmentation. The deep learning based model is extensively utilized due to its initiation from the much prevailing impression of the neocortex. Recent advancements [19] in this area have validated and confirmed its applications in enormous parts.

This effort offers a framework that empowers to take complete advantage of both contextual and spectral characteristics present in the hyperspectral scene by mining the features at both ends, i.e., mining spectral characteristics through deep learning rooted auto-encoder [38], and mining contextual characteristics through HMRF [5]-based segmentation which results in a powerful SAE-HMRF-based feature extractor for accurate performance. In order to enhance the classification capability of the prescribed approach, merging of both of these spectral and contextual neighboring characteristics is on the rise [19, 45, 46]. In said approaches, the association of a particular HSI pixel to a class is decided on two factors, i.e., one is pixel rooted classification outcome and the second is the contextual-based segmentation. Therefore, an appropriate segmentation technique can enhance the quality of the classification manifold. The developed technique in this section of the chapter follows these two vital steps. First, it classifies the scene based on the pixel rooted classification by utilizing deep learning-based Sacked auto-encoder architecture for characteristic mining. Secondly, spatial characteristics are mined by utilizing the flexible edge detection-based region growing segmentation process approved by HMRF-EM [5]. After mining these two types of features, the majority vote (MV) process is deployed to fuse the spectral and contextual results (Fig. 6.5).

6.3 Proposed Methodology

The overall working of the developed approach is depicted in Fig. 6.8. Figure 6.9 displays the complete structure for classification. A precise description of each concept is explained in the following sub-parts of the chapter.

6.3.1 HMRF-EM Segmentation

In order to perform the accurate segmented regions of the hyperspectral image, it is effective to take the first component of principal component analysis $y = (y_1, y_2, \ldots y_n)^T$. Let's suppose the HSI scene H comprising a resolution $R \times C$.

I. Raw SAE Pixel-wise classifier

Input: hyperspectral dataset

Output: classification map

1. Preprocessing stage.
2. Training stage.
3. Testing stage.
4. Validation stage.

II. HMRF Segmentation

Input: hyperspectral image

Output: single band segmented map

1. Dimension reduction using PCA
2. Gradient calculation
3. K-means processing
4. HMRF-EM Segmentation

III. Combination of Maps

Input: SAE classification map, segmentation map

Output: spectral – spatial classification map

1. Voting stage
2. Winner stage
3. Updating stage

Fig. 6.8 Spectral–spatial classification process. I Spectral step; II Spatial step; III Combination step

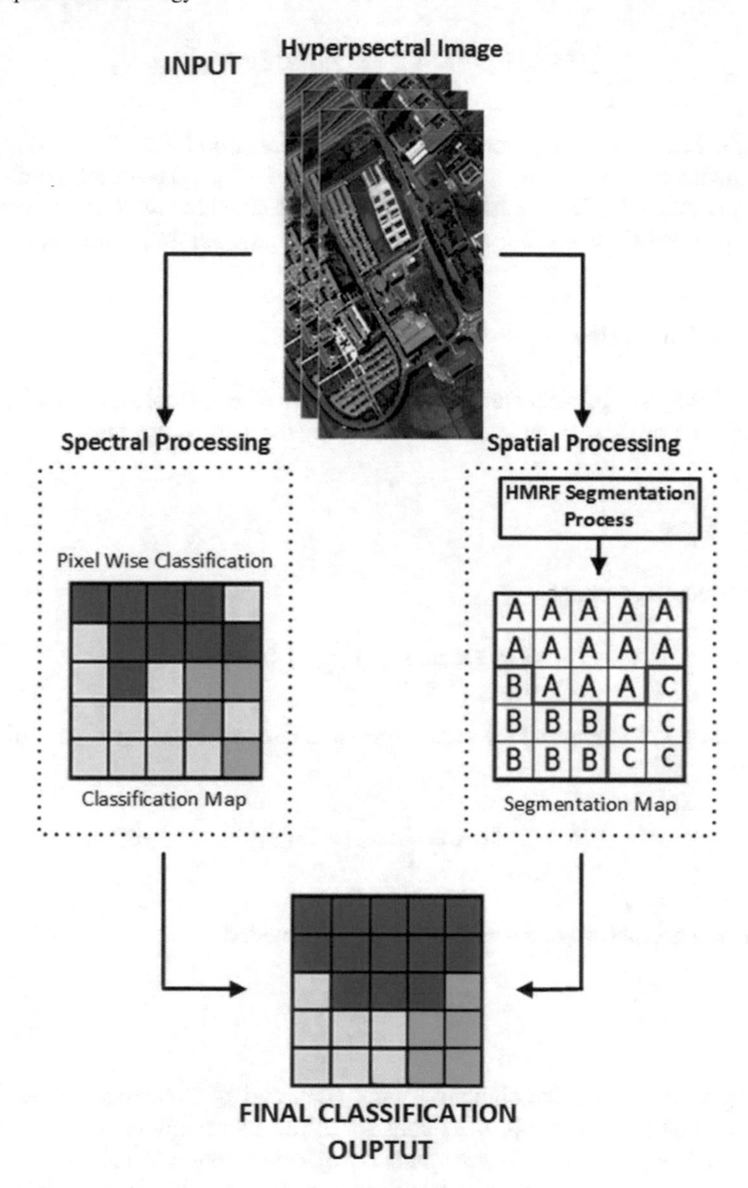

Fig. 6.9 Proposed scheme for spectral–spatial classification

The hidden Markov random field can be shaped as

$$p(x, y; \theta) = f(x) \prod_{b=1}^{N} p(y_b|a_b) \tag{6.8}$$

$$p(y_b|x_{N_b};\theta) = \sum_{k \in K} g(y_b;\theta_l)q(k|x_{N_k}) \tag{6.9}$$

where $f(x)$ expresses the probability distribution function (pdf) of x, $g(y_b;\theta_k)$ is a pdf consisting of $\theta_k = (\mu_k, \sigma_k^2), a = (a_1, ...a_N)^T$ and $q(k|a_{N_k})$ represents probability mass function for H. The method of fine-tuning the suitable architecture comprises 3 initial phases [47] which are explained in detail in the coming section.

6.3.1.1 Initialization

ISODATA [48] The primary starting parameters of $x^{(0)}$ and $^{(0)}$ are generated through the ISODATA as these parameters are utilized by the same procedure.

6.3.1.2 MAP

MAP procedure comprises

$$\hat{x} = \arg\max_{x \in B}\{p(y|x;\theta)f(x)\} \tag{6.10}$$

Here, $p(y|x,\theta)$ represent the same expression as depicted in equation 1 and $f(x)$ is provided by

$$f(x) = \frac{1}{P}exp(-W(x))$$

Here, P is a normalizing constant and $W(x)$ is provided by

$$W(x) = \sum_{Q \in q} R_q(x) \tag{6.11}$$

Here, $R_q(x)$ represents the clique potential (CP) and Q expresses a collection of all cliques that are feasible [49]. Q consists of the subsection of the pixel. In this subsection, every couple of distinct parts is adjacent to one another [5], exempting individual site cliques. In the recent research, adjacent 4-connected neighbor was taken into account. Therefore, CP is computed as

$$Q_q(a_i, a_j) = \frac{1}{2}(1 - H_{x_i,x_j}) \tag{6.12}$$

MAP can also be acknowledged as an energy function to minimize

$$\hat{x} = \arg\min_{x \in B}\{W(y|x) + W(x)\} \tag{6.13}$$

where $W(y|x) = \sum_j [\frac{(y_j - \mu x_j)^2}{2\sigma_{a_j}^2} + \frac{1}{2} log\sigma_{y_j}^2]$. $W(y|x)$ computes the fitting. On the other hand, the next part is utilized for contextual regularity. Several great accurate processes are feasible and possible for finding the solution of MAP. We exploited the process depicted in [47].

6.3.1.3 Expectation Maximization Algorithm

This process is exploited to estimate the factor θ. Initially, the "Expectation" phase is estimated with the following equation:

$$P(\theta|\theta^{(H)}) = Exp\left\{\log p(y, x; \theta)|y, \theta^{(H)}\right\} \tag{6.14}$$

This equation is maximized in order to compute the next parameter [48].

6.3.2 *Final Segmentation with Preserved Edges*

Hidden Markov random field may end up having over-regularity within the scene. So in order to address this issue to sustain the original image boundaries during this process of segmenting the scene, a dimensional reduction procedure is employed through Principal Component Analysis and the initial 4 components are engaged due to the presence of maximum variational information in the specified selected components. Moreover, in order to have an enhanced effective boundary, a Sobel edge detector is computed on the specified PCs. The summation is then applied to the outcome of the Sobel step. The resulting segmentation outcome is transferred to binary form which is the final form of a binary segmentation map.

6.3.3 *SAE Pixel-Wise Classification*

As mention in previous chapters, stacked auto-encoders (SAE) is a deep learning-based architecture that is utilized for prominent, distinct, and deep mining of features that exist in complex data. This DL model has already proved its powerful feature mining capability in the major fields of computer vision, signal processing, and speech recognition. SAE was selected in this section for feature mining due to its very strong data mining capability particularly in HSI as has been demonstrated by previous works [24]. The SAE DL model comprises several encoders. An encoder is a one-layer DL model that comprises an input, a hidden, and an output layer, as depicted in Fig. 6.10 taken from [38]. The responsibility of the encoder is to first encode the given input information with the help of parameters and hidden neurons and then as a next step, regenerate the same information in such a way that near to original

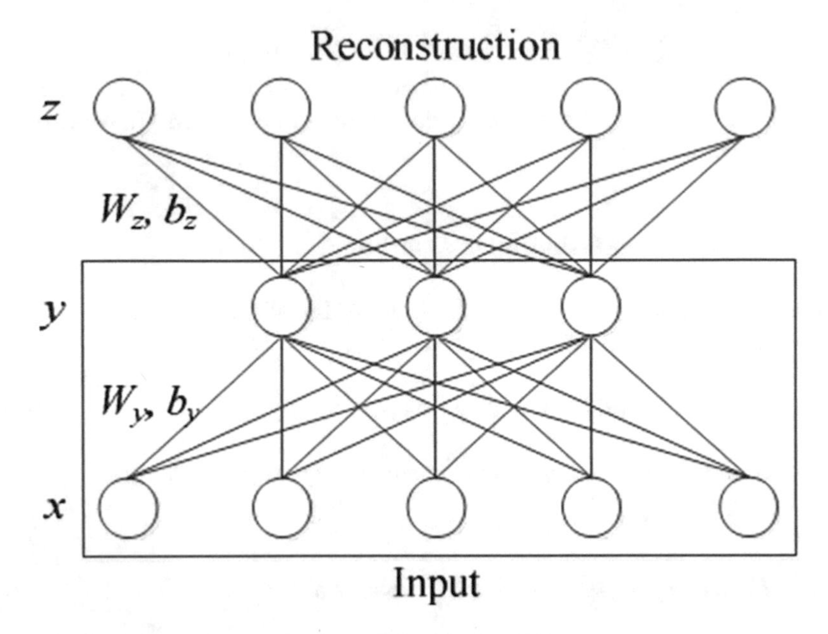

Fig. 6.10 Single layer auto-encoder for HSI classification [38]

data is regenerated; this part is called decoding in the DL world. In this process, the model tries to learn the values through which it can regenerate the data and hence it learns the distinguishing characteristics of the data that exists in that particular information. Multiple numbers of this encoder are placed over one another to form a stack of encoders. These several encoders are required for characteristic mining if the given data is complex in nature. This phenomenon clearly suits the complex and multi-dimensional nature of the hyperspectral scene as the scene comprises hundreds of spectral. This topic can be read in detail from [50]. This work is rooted from [38] to mine the features of HSI for subsequent classification.

This procedure can be written in mathematics language as

$$\phi = \sigma_1(w_\phi x + b_\phi), \psi = \sigma_2(w_\psi y + b_\psi) \tag{6.15}$$

where w_ϕ, w_ψ represents the weights and $\sigma_1(.)$, $\sigma_2(.)$ denotes the sigmoid activation function. One of the major objectives is to reduce the regeneration error.

Effective cost function is required for the whole process. This was accomplished through cross entropy rooted CF [38]. In addition to CF, sigmoid rooted activation is considered. The CF and activation function can be depicted mathematically as

$$CF = -\frac{1}{c}\sum_{i=1}^{d}\sum_{j=1}^{m}[x_{ij}\log(\psi_{ij}) + (1 - x_{ij})\log(1 - \psi_{ij})] \tag{6.16}$$

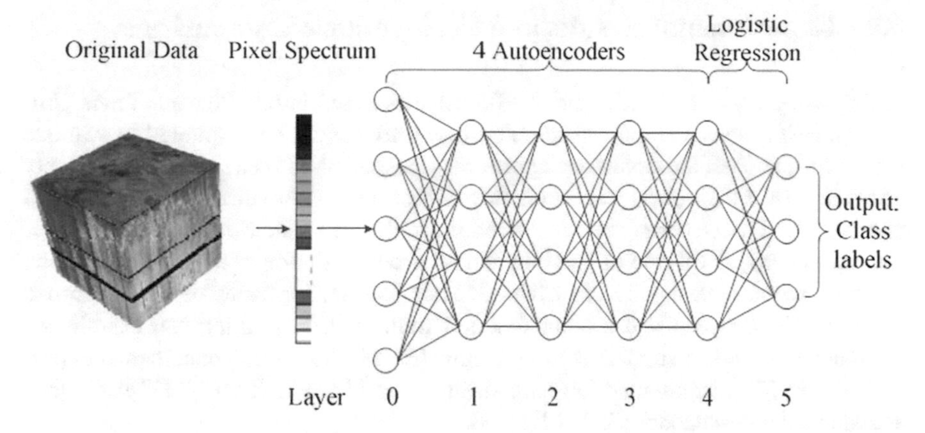

Fig. 6.11 Deep spectral representation with 1 input and output layer each, 4 hidden layers, and one output layer of Logistic regression [38]

Here, m and c show height of a vector and volume of a mini-batch correspondingly. Moreover, $x_{ij}(\psi_{ij})$ shows the specified component of a particular input. To optimize the problem, mini-batch stochastic gradient descent is exploited to resolve the equation.

A comprehensive study of the stacked auto-encoder from an HSI perspective along with its details from an implantation point of view can be found in [38]. Figure 6.11 received from [38] clearly depicts the processing of SAE including its layer-wise description. After mining the features through the layer-wise model, logistic regression is employed for ultimate pixel-based classification.

6.3.4 *Majority Voting*

At this point, the output of two different algorithms for HSI classification is obtained. The output of the HMRF approach contains the binary segmentation map of the scene while the output of the SAE deep learning-based classification approach contains the pixel-based classification results of the scene. As a final step, some methodology must be employed to fuse the result from both the approaches so that each approach can overcome the weakness of the other approach. For this, the majority voting (MV) rooted technique is adopted as described by authors in [51] also depicted in Fig. 6.9. In order to perform majority voting on the outcome of each technique, the volume of pixels that belong to a particular class for each technique are counted. Afterwards, all the pixels are given under the ownership of a particular class if most of the pixels belong to that class. Hence, following this way, the strongest point of each technique that belongs to a totally different category of algorithms gives their best result and the most accurate participation of each technique is contributed at the end.

6.4 Experimental Results and Performance Comparisons

In this work, regular AVIRIS and ROSIS sensors based Indian Pine and Pavia University, HSI scenes that are considered difficult to classify are exploited to evaluate the performance of the developed approach. The described scenes have been widely utilized all over the world by researchers to access the performance of their developed approaches. These scenes are considered difficult due to their obscurity, multiplicity, and disrupted volume of existing sampled data. A comprehensive explanation of every scene is designated in Chap. 2, Sect. 2.3. Performance of the developed technique is analyzed and a comparison is made with prevailing and established approaches including support vector machine (SVM) [33], orthogonal matching pursuit (OMP) [52], augmented Lagrangian-multilevel logistic (LORSAL-MLL), [53] and stacked auto-encoder (SAE-LR) [38].

As explained before, as the first step dimensional of the provided HSI scene are reduced through principal components analysis, and initial 4 components are engaged due to the presence of high variational data in them. Afterwards, the Sobel boundary detection approach is computed on each PCA component and the average of the 4 components is employed. Finally, the output is changed to the binary form of the scene to compute the slope as depicted in [5]. The given scene is computed by ISODATA and the outcome of this step and the preceding phase are normalized through HMRF-EM [5].

Spectral characteristics are also mined and pixel-based classification is computed correspondingly by utilizing stacked auto-encoder-based logistic regression. Finally, the outcome of both the approaches, i.e., contextual segmented data and spectral pixel-based classified outcome are fused together by utilizing the majority voting criteria. The outcome of these phases is presented in Figs. 6.12 and 6.13.

Figure 6.13 and Tables 6.4 and 6.5 clearly present the improved performance of the developed SAE-HMRF methodology in comparison to the SAE-LR technique unaccompanied and with even enhanced performance in comparison to other established approaches by also considering the contextual characteristics. It is also evident

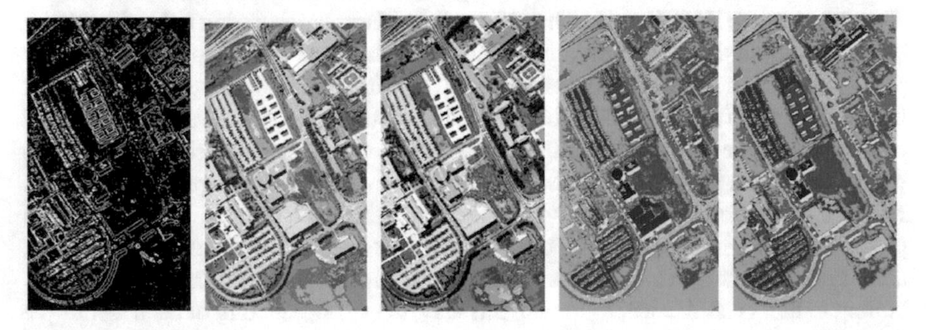

Fig. 6.12 Spectral–spatial phases of Pavia University dataset **a** edge detection phase **b** ISODATA **c** HMRF-EM **d** SAE-LR-based classification **e** Proposed SAE-HMRF-based classification

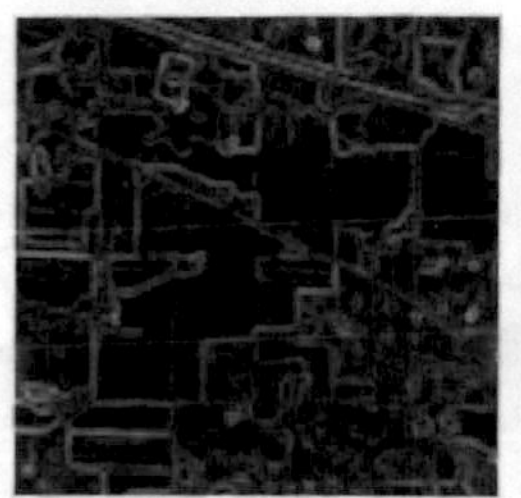 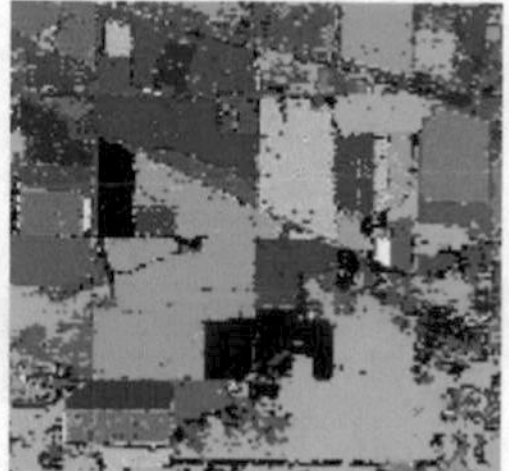 

Fig. 6.13 Spectral–spatial classification of the Indian Pine dataset **a** boundary detection phase **b** SAE-LR-based classification **c** Proposed SAE-HMRF-based classification

Table 6.4 Classification accuracies of each class for the Indian Pine image using 9% training samples obtained by SVM [33], OMP [52], LORSAL-MLL [53], SAE-LR [38], and SAE-HMRF

Class	Train	Test	SVM	OMP	LORSAL-MLL	SAE-LR	SAE-HMRF
1	5	49	77.73	55.12	**100.0**	93.33	94.11
2	143	1291	77.35	61.60	87.46	84.66	**89.42**
3	83	751	78.56	58.62	81.23	84.39	**88.70**
4	23	211	68.75	42.21	88.41	73.08	**88.51**
5	50	447	88.87	87.29	**97.49**	93.47	94.54
6	75	672	89.12	95.30	**97.34**	93.41	95.31
7	3	23	95.37	85.20	**100.0**	**100.0**	**100.0**
8	49	440	95.09	96.44	**97.98**	95.11	96.01
9	2	18	67.65	36.67	**100.0**	**100.0**	**100.0**
10	97	871	78.64	71.10	83.64	85.78	**90.73**
11	247	2221	81.19	74.11	86.31	83.46	**89.11**
12	61	553	79.74	51.05	**91.79**	81.62	85.27
13	21	191	92.26	96.85	**100**	98.52	**98.87**
14	129	1165	92.72	91.85	**97.04**	91.77	93.45
15	38	342		69.79	41.67	86.18	**89.95**
16	10	85	97.96	91.90	**100.0**	97.96	**99.05**
Overall Accuracy			82.91	73.38	87.18	86.85	**90.08**
Average Accuracy			83.17	71.06	90.83	89.95	**93.09**
Kappa Coefficient			0.805	0.696	0.8536	0.8495	**0.8875**

from the outcome provided result in Tables 6.5 and 6.4 that the developed approach also has shown improved performance in overall accuracy calculation and kappa constant.

Table 6.5 Classification accuracy of each class for the Pavia University image using 9% training samples obtained by SVM [33], OMP [52], LORSAL-MLL [53], SAE-LR [38], and SAE-HMRF

Class	Train	Test	SVM	OMP	LORSAL-MLL	SAE-LR	SAE-HMRF
1	597	6034	97.50	64.16	**100.0**	96.88	97.18
2	1681	16971	97.70	82.23	98.50	98.30	**99.01**
3	189	1910	78.53	71.04	**96.27**	91.09	93.50
4	276	2788	89.29	93.43	91.03	99.15	**99.40**
5	121	1224	98.77	**99.90**	98.45	99.85	99.89
6	453	4576	83.04	69.47	97.24	96.44	**97.51**
7	120	1210	64.58	87.31	97.94	94.12	**95.96**
8	331	3351	86.90	71.57	99.94	93.27	**95.14**
9	85	862	99.92	97.27	**100.0**	**100.0**	**100.0**
Overall Accuracy			92.04	78.07	98.59	97.12	**98.72**
Average Accuracy			88.47	81.82	97.28	96.56	**97.46**
Kappa Coefficient			0.963	0.711	0.9719	0.9615	**0.9746**

6.5 Summary of the Proposed Hyperspectral Image Classification Method Based on Deep Auto-encoder and Hidden Markov Random Field

A new classification framework, established on SAE-based pixel-wise classification and HMRF-based segmentation, is proposed. This approach is sensitive to both spectral and spatial features of the given scene. In the last stage, the majority voting criteria is exploited to integrate the extracted pixel-wise result and HMRF-based segmentation result. Presented experimental results on AVIRIS and ROSIS-based datasets confirm the effectiveness of the designed framework.

6.6 Hyperspectral Image Classification Based on Hyper-segmentation and Deep Belief Network

Successful and exact hyperspectral image (HSI) investigation has gotten increasingly significant with headways in remote detecting innovation, procuring information with several spectral channels alongside the definite spatial data of the scene [6]. The rich spectral and spatial data if effectively used can create higher order correctness [1]. Besides, ongoing advances in remote detecting sensors have empowered to procure rich spatial goals [2], henceforth made it conceivable to assess the little spatial land covers in the HSI. Because of the natural rich spectral–spatial data, HSI characterization has been a functioning examination region. Be that as it may, it likewise acts colossal difficulties such as higher dimensionality, constrained marked sample, Hughes phenomenon [14], and heterogeneity [54].

A huge number of spectral channels yet predetermined number of tests samples prompts revile of dimensionality [14].Therefore, exploring spatial/auxiliary characteristics alongside the spectral characteristics and structure of classifiers assumes a pivotal job in HSI characterization. Numerous strategies have been developed so far to manage the HSI characterization. Conventional pixel-based classifiers take every pixel self-sufficiently without considering spatial data. K-nearest neighbor classifier (K-NN), restrictive arbitrary fields [55], neural systems [56], and support vector machine [33, 57] have been researched. Out of these pixel-based classifiers, SVM accomplished improved results because of its capacity to deal with high-dimensional information. A dominant part of the previously mentioned classifiers experiences reviles of dimensionality and restricted test information [29]. Besides, spatial data isn't considered, as there is a solid relationship between adjoining pixels [6]; however, pixel-wise classifiers take every pixel freely. Dimensionality decrease approaches were likewise proposed to deal with the higher dimensionality and restricted preparing tests. Principal component analysis (PCA) [58] and free component analysis (ICA) [59] are a portion of the notable methodologies. These methodologies shrink the spectral measurements from hundreds to condense the information into only a few bands (2 or 3) which subsequently ends up harming the spectral data. Band choice determination is another method of dealing with the previously mentioned issues.

The fusion of contextual characteristics along with spectral characteristics to attain the much needed enhanced classification accuracy is one of the realities that got attention in the current years [60]. In recent years, it has been strongly realized the importance of contextual data in HSI due to its contribution in the subject task, additionally the HSI acquiring sensors has also incorporated instruments that have the capacity to collect the scene which is not only rich in spectral domain but also contains a respectable high resolution hence having an improved spatial domain as well. Hence, now it is the reality that considering both spatial and spectral characteristics ends up having enhanced classification outcome [23]. Hence, considering contextual association is a necessity now.

Approaches based on spectral and contextual characteristics can be distributed into 3 groups which include the incorporation of contextual data apart from spectral characteristics. The first group includes techniques based on the fusion of contextual data prior to classification, the second group considers fusion at the time of classification procedure while in the third group, fusion is considered post classification. In the first group, extraction and integration of contextual characteristics are performed prior to the classification; these include contextual characteristic mining using morphological profiles [61–63] and also with the aid of flexibly detecting the actual edges [23]. Likewise, multiple kernel-based approaches integrate contextual characteristics along with spectral properties [64, 65]. Conversely, these contextual feature mining approaches involve manual tuning and information and are typically handcrafted. The second group of classification methods includes the addition of contextual characteristics into a classifying technique through a classification procedure;

it includes methods such as statistically learnable theory (SLT) [54] and concurrent subspace detection (SSP) [66]. The third group comprises approaches that add the contextual weightage after the independent classification procedure. For instance, researchers in [67] initially applied support vector machine for pixel-based spectral classification. Afterwards, watershed segmenting is applied for contextual properties mining tailed by majority voting which is applied on the pixel-based spectral results and segmentation results to have the final output. For instance, researchers in [53] applied augmented Lagrangian multilevel logistic with a multilevel logistic (MLL) prior (LORSAL-MLL). Likewise, researchers in [4] fused the outcome from the segmenting procedure and stacked auto-encoder rooted classification with the aid of majority voting.

Lately, in the past few years, neural network has put a revolutionary impact, and DL has demonstrated its efficacy and worth in numerous areas especially in computer visualization such as scene classification [68], communication identification [69], and linguistic handling [70]. Deep learning rooted models have significant contribution in hyperspectral scene analyses too [71]. Nevertheless, including contextual influence into a DL model is quite a significant issue.

In this part of the chapter, a new approach is developed and explained which is based on the DL rooted deep belief network (DBN)-based architecture for spectral characteristic mining and flexible edge detection-based contextual characteristic mining; these two spectral and contextual outcomes are fused at the end for proficient and improved classification performance. Spectral characteristic mining is performed with the aid of the DL rooted DBN model [72] along with logistic regression (LR) which is utilized as a pixel rooted classifier. On the other hand, contextual characteristic mining is done by employing flexible edge evaluation rooted segmentation which was already established in Chap. 4. This developed procedure falls in the third group of classification approaches where the spectral and contextual characteristics are combined at the end. The final allocation of a particular pixel to a specific class depends on both, the weightage of spectral classification performed through DBN and the weightage of contextual characteristic mining done through the boundary detection-based segmentation method.

6.6.1 Proposed Methodology

It is sturdily understood in HSI research community that joining contextual characteristics can meaningfully increase the algorithm performance [73]. The developed technique initially explores multiple-layer deep belief net for spectral characteristic mining and Multiple Logistic Regression is employed for succeeding pixel-based classification. To obtain the meaningful contextual regions of the scene, flexible edge detection rooted segmentation is employed. In the third stage, majority elective [74] rooted procedure is applied to fully activate and combine the spectral and contextual characteristics for ultimate spectral–contextual classification. A meticulous explanation of every stage is illustrated in Figs. 6.14, 6.15, and 6.16.

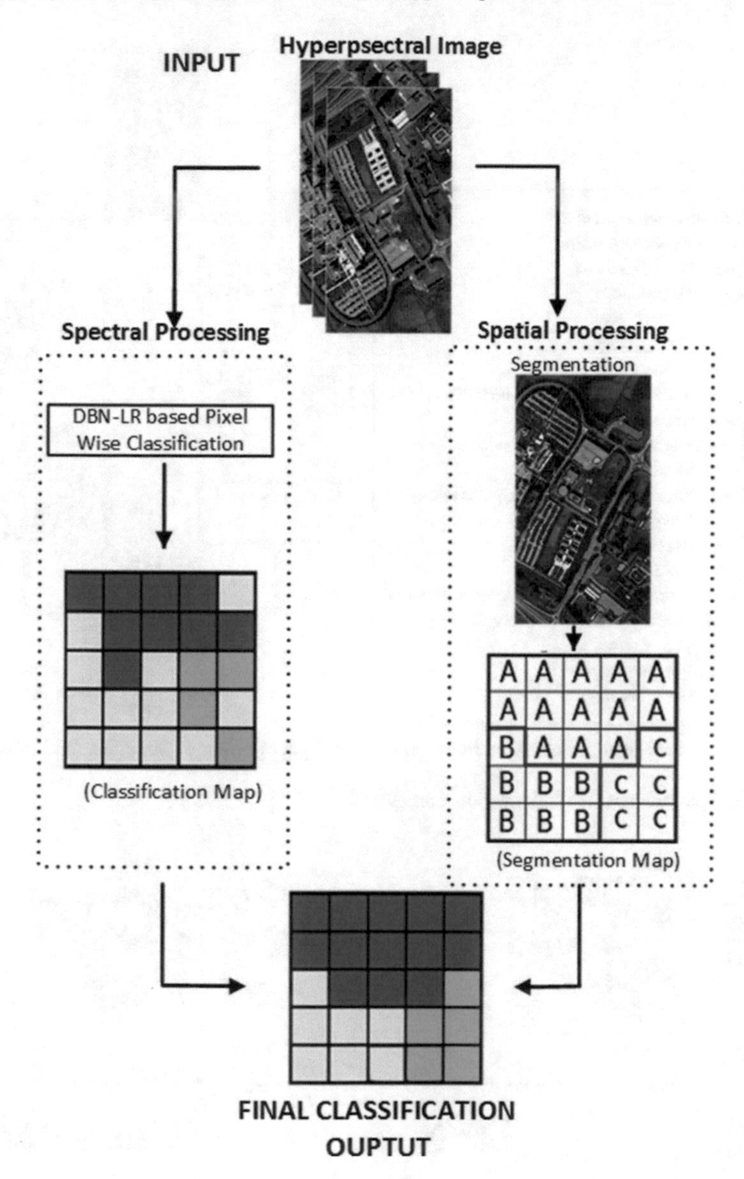

Fig. 6.14 Spectral–spatial classification framework

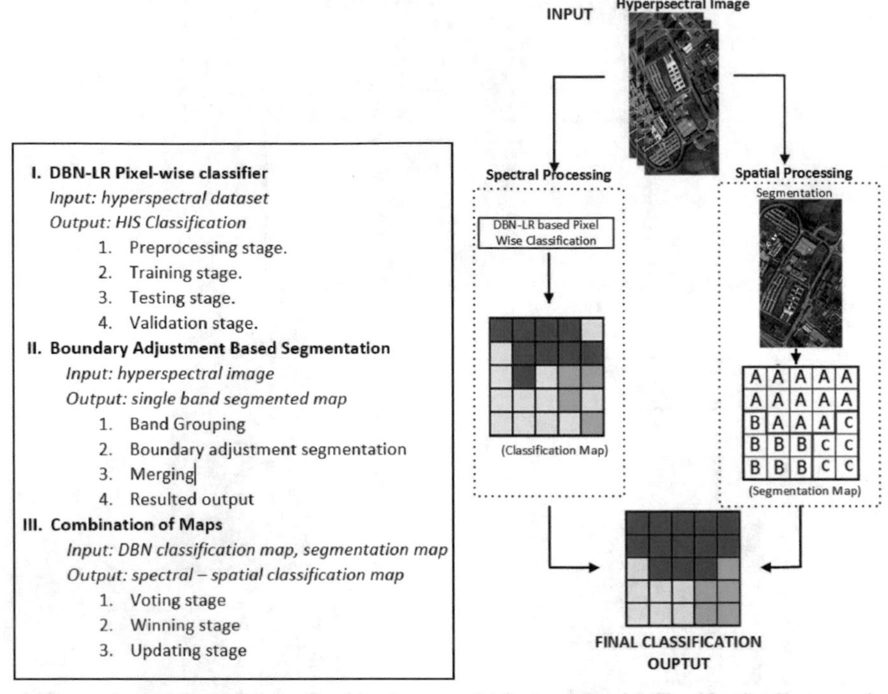

(a)Spectral stage, Spatial stage, Combined stage (b) Spectral-Spatial Classification Framework

Fig. 6.15 Spectral–spatial classification framework

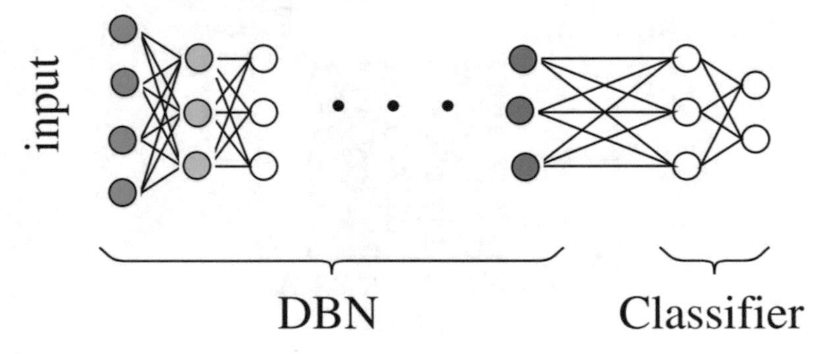

Fig. 6.16 Framework of the DBN-based pixel-wise classification

6.6.1.1 Spectral Feature Extraction via DBN

Deep belief net is comprised of the neural network (NN) rooted restricted Boltzmann machine (RBM) learnable model that contains first the visible layer which is used to take the input raw spectral data x, then comes the next layer y which learns to save the distinguishing characteristics through parameters with higher association in the provided spectral information as presented in Fig. 6.15. The overall functional form can be explained as

$$E(x, y, \theta) = -\sum_{j=1}^{m} \frac{(x_j - b_j)^2}{2\sigma^2} - \sum_{i=1}^{n} a_i y_i - \sum_{j=1}^{m}\sum_{i=1}^{n} w_{ij} \frac{x_j}{\sigma_i} y_i \tag{6.17}$$

The restrictive distributions are provided by

$$P(y_j|x; \theta) = h\left(\sum_{i=1}^{m} w_{ji} x_j + a_i\right) \tag{6.18}$$

$$P(x_j|y; \theta) = V\,distributions\left(\sum_{j=1}^{m} w_{ji} y_j \sigma_j^2 + b_j\right) \tag{6.19}$$

Here, σ depicts the SD of a Gaussian observable part, and $V(.)$ is the Gaussian distribution.

DBN mostly includes a restricted Boltzmann machine that is placed over each other while the training of RBM contributes as a critical part in DBN. At the learning phase, available sampled data is given in order to get the whole system trained and to learn the given parameters to get the contextual and spectral data from the HSI scene. This process proceeds with the fine-tuning of parameters that involves the back propagation procedure. At the classification phase, a trained net model is utilized to apply the classification for a given data, hence it results in independent pixel-based classification results. Here, in this procedure DBN-LR is applied which uses a deep belief net for characteristic mining from spectral information of the scene.

6.6.1.2 Spatial Feature Extraction via Hyper-segmentation

In the spatial domain, the research community heavily believed in 2 major conditions which play a major role during contextual feature mining as already explained in previous chapters. The first conditional statements explain that pixel values whose spectral characteristics are similar may share the identical class with great possibility. Secondly, pixels that are in neighbors and have great associativity from a spectral point of view may also share the identical final class. Therefore, to take these proper-

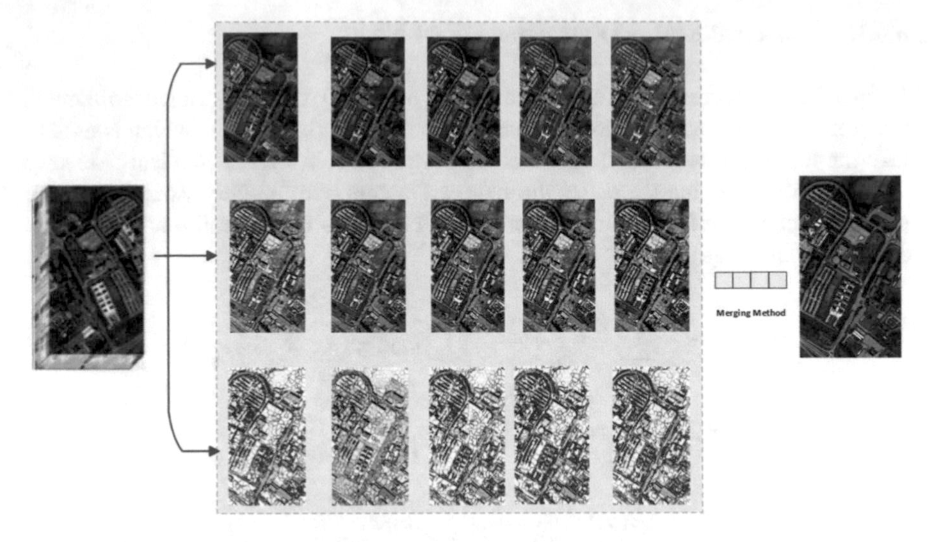

Fig. 6.17 Framework of the hyper-segmentation process

ties of contextual association into account, the developed flexible boundary detection procedure is developed and applied. This approach is already explained in greater perspective in Chap. 4.

There is a balancing equation that contributes critically and comprises 3 major factors and can be mathematically modeled as

$$A(b, a_i) = \sqrt{|x_b - m_i|^2 + \lambda \tilde{n}_i(b)|Grad(b)|} \tag{6.20}$$

Here, x_b is said to be a spectral vector which exists for each edge pixel, m_i is the mainstream vector, $\tilde{n}_i(b)$ is called the straightening function, and $|Grad(b)|$ expresses the slope at marked pixel b. The complete execution of the approach for contextual segmentation is described in [28].

6.6.1.3 Majority Voting

The respective outcome received separately from DBN-LR (spectral-based classification) and the contextual outcome received from the flexible edge-based approach are combined through mainstream voting [74] so that the maximum weightage of each outcome can be accommodated to form the final classification map. In the said outcome fusion approach, every pixel in resulting segmented regions is allocated to a maximum iterative class allotted by the DBN-LR classifier. In this way, maximum response from each outcome can be accommodated to have a more improved classification map (Fig. 6.17).

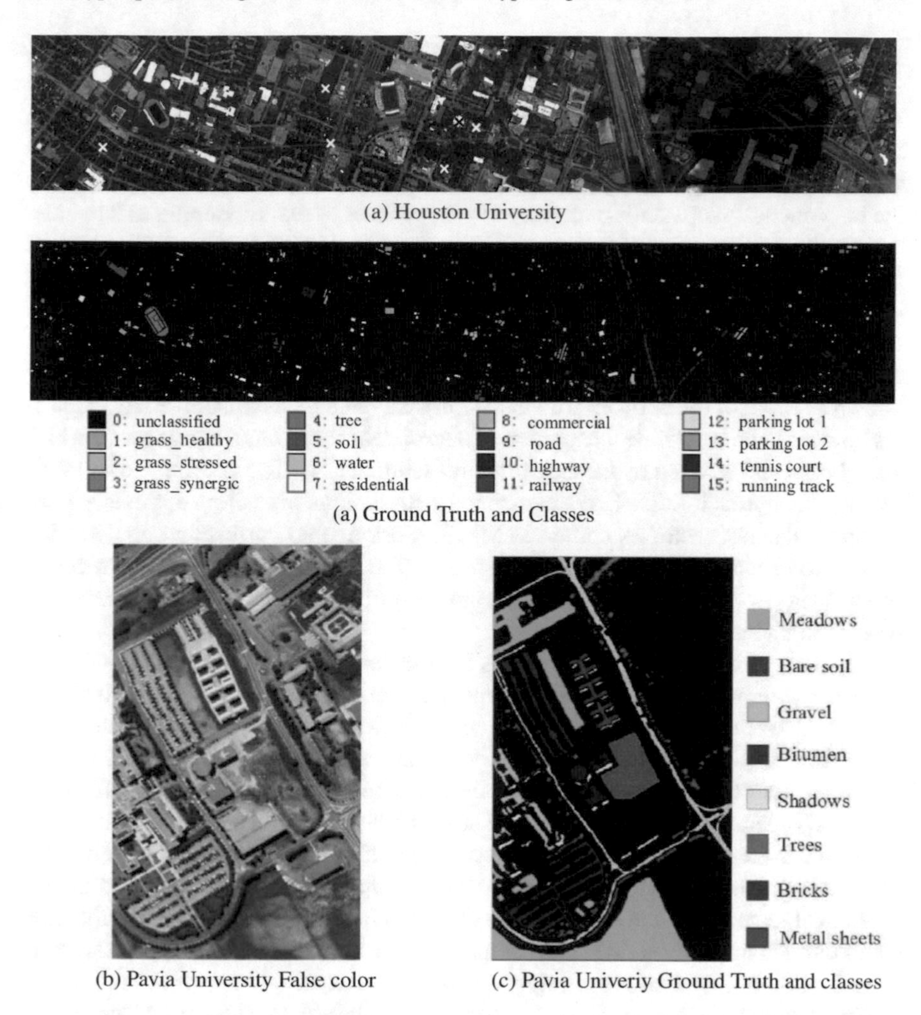

Fig. 6.18 Hyperspectral Image datasets

6.6.2 Experimental Results and Performance Comparison

To evaluate the execution and output of the developed methodology, various experiments are conducted on famous, difficult, and complex AVIRIS and ROSIS sensor based Houston University and Pavia University scenes as shown in Fig. 6.18. These scenes are extensively applied by most of the researchers all over the world to demonstrate the outcome of their algorithms. A meticulous explanation of each scene is depicted in Chap. 2, Sect. 2.3.

6.6.2.1 Spectral–Spatial DBN-HS Classification

The experimental setups that were utilized comprised a Windows 10 system, with 3.0 GHz computer and graphics card of specs NVIDIA GeForce GTX 980. The implantation coding language utilized was Theano. The quantity of inner layers in the network has a substantial weightage in the process of feature mining as it decides the quality and quantity of feature mining which in turn affect the performance of the classification. But it doesn't employ that increasing the number of inner layers increases the performance, i.e., performance is not directly proportional to the number of hidden layers. For every HSI data scene, only 10% of the existing labeled data was randomly selected and applied for training proposes. The number of inner layers also known as depth of the network for Pavia University and Houston University datasets was chosen to be two while the number of neurons for every inner layer was 50 which was chosen and advised by the experimental setup in [24]. The accuracy level of the developed approach DBN-HS is associated with famous prevailing approaches, for instance, support vector machine (SVM) [33], orthogonal corresponding detection (OMP) [52], and deep belief net through LR (DBN-LR) [24] along with recently established deep CNN (CNN) [35]. In the case of DBN-LR, spectral information was taken into account.

Separate performances of every class for the developed approach along with the present techniques for the described hyperspectral scenes are depicted in Table 6.6 and 6.7. One of the vital difficulties and complexities faced in these HSI scenes were firstly the size of the smallest object in the scene, secondly the noise, and thirdly the amount of assorted pixel which is due to the small contextual features. In the Houston University scene, the developed methodology acted really well specifically for structures with the smallest contextual size. And we believe, it is because of the effective flexible boundary detection-based segmentation approach that was applied during contextual feature mining, and later its effect was included in the final classification result through majority voting. The comprehensive picture of the hyperspectral scene classification outcome of the developed approach is depicted in Fig. 6.19.

Every color shown in the scene corresponds to a particular category of the earth's surface region which is exactly similar to the ground truth scene mentioned before. The final performance as demonstrated approves the developed approach both conceptually and practically as demonstrated through experiments that contextual features also play a vital role in improved performance. Moreover, contextual characteristics also aid in removing and overcoming the salt and paper noise.

Generally, the investigational outcome establishes the important development in hyperspectral scene classification by uniting contextual data and spectral characteristic mixture. The developed algorithm has accomplished meaningfully on small contextual-sized regions.

Table 6.6 Classification accuracy (%) of each class for the Houston University dataset obtained by SVM [33], OMP [52], and CNN [35] using 10% training samples

Class	Training	Test	SVM	OMP	CNN	DBN-LR	DBN-HS
1	125	1126	97.47	98.27	81.20	**99.20**	99.0
2	125	1129	98.32	98.10	83.55	**99.60**	99.20
3	70	627	99.37	99.68	99.41	**100.0**	**100**
4	124	1120	98.01	96.70	91.57	**99.60**	**99.60**
5	124	1118	96.01	98.48	94.79	99.60	**99.60**
6	33	292	99.83	97.95	95.10	97.2	**98.05**
7	127	1141	91.23	86.90	63.53	97.0	**98.16**
8	124	1120	86.23	89.82	42.64	97.8	**98.0**
9	125	1127	86.99	79.37	58.17	94.0	**95.25**
10	123	1104	91.42	89.68	41.80	97.4	**97.95**
11	124	1111	91.67	82.77	75.71	97.3	**98.1**
12	123	1110	87.05	81.94	84.15	95.2	**96.26**
13	47	422	78.16	35.55	40.00	88.0	**91.1**
14	43	385	97.42	98.18	98.79	**100**	**100**
15	66	594	99.49	98.40	97.89	**100**	**100**
Overall Accuracy			93.06	89.70	85.42	97.70	**98.98**
Average Accuracy			93.25	88.78	76.55	97.50	**98.46**
Kappa Coefficient			0.925	0.889	0.7200	0.975	**0.9875**

Table 6.7 Classification accuracy of each class for the Pavia University dataset obtained by SVM [33], OMP [52], and CNN [35] using 10% training samples

Class	Training	Test	SVM	OMP	CNN	DBN-LR	DBN-HS
1	597	6034	97.50	64.16	87.34	87.37	**89.78**
2	1681	16971	97.70	82.23	**94.63**	92.10	94.01
3	189	1910	78.53	71.04	**86.47**	85.57	**88.50**
4	276	2788	89.29	93.43	96.29	95.11	**97.40**
5	121	1224	98.77	99.90	99.65	99.74	**99.89**
6	453	4576	83.04	69.47	93.23	91.94	**94.30**
7	120	1210	64.58	87.31	**93.19**	92.21	93.96
8	331	3351	86.90	71.57	86.42	87.02	**88.14**
9	85	862	99.92	97.27	**100**	**100**	**100**
Overall Accuracy			92.04	78.07	92.56	91.18	**93.98**
Average Accuracy			88.47	81.82	93.02	92.34	**93.46**
Kappa Coefficient			0.903	0.711	0.9006	0.8828	**0.9175**

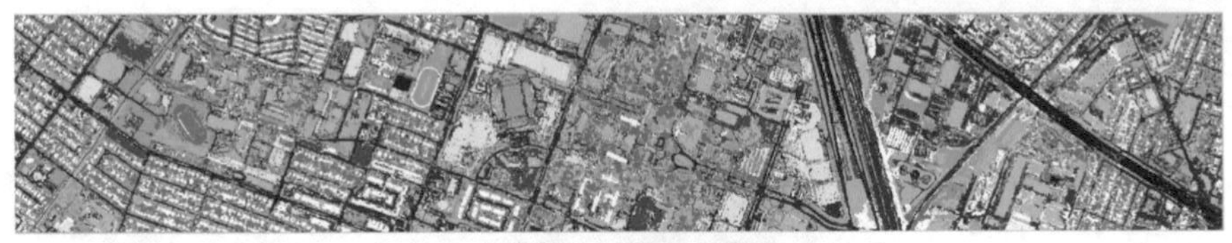

(a) Houston University

(b) Pavia University

Fig. 6.19 Classification results of Houston University and Pavia University datasets using the proposed method

6.7 Summary of the Proposed Deep Learning-Based Methods for Hyperspectral Image Classification

This chapter covers deep learning (DL)-based methods for hyperspectral image (HSI) classification where spatial features are incorporated after classification.

In the first proposed method, segmented deep belief net (DBN) is exploited, where spectral channels are segmented and grouped into similar bands, and DBN is applied on each group of bands separately, resulting in more robust and abstract features. Managing plenty of characteristics simultaneously results in complication and intricacy and damages the accuracy of DBN. Therefore, the dual-phase classification method is developed by fusing the characteristics of contextual properties through segmentation and spectral segmentation, partitioning the actual spectral area into several associated channels, and then employing DBN distinctly to every segment. Therefore, dropping the intricacy of the learning procedure and mining native characteristics makes it easier for DBN to effectually mine the spectral–spatial characteristics.

In the second part of the chapter, DL-based techniques are exploited where spatial contextual features are incorporated after the classification stage, resulting in better performance. Two main DL architectures stacked auto-encoder (SAE) and DBN are used for this purpose. First, DL architectures are exploited to extract abstract and representative features. Second, the spatial segmentation process is carried out by the HMRF-EM-based technique and through the proposed segmentation technique. Finally, the majority vote (MV) procedure is employed to integrate the spectral and spatial outcomes that lead to the final classification results.

Extensive experimental analysis was carried out on AVIRIS and ROSIS sensors-based diverse HSI datasets. The proposed DL-based approaches produced better classification results and outperform existing well-known and widely used HSI classification approaches.

References

1. Fauvel M, Benediktsson JA, Chanussot J, Sveinsson JR (2008) Spectral and spatial classification of hyperspectral data using svms and morphological profiles. IEEE Trans Geosci Remote Sens 46(11):3804–3814
2. Ghamisi P, Dalla Mura M, Benediktsson JA (2015) A survey on spectral–spatial classification techniques based on attribute profiles. IEEE Trans Geosci Remote Sens 53(5):2335–2353
3. Ghamisi P, Couceiro MS, Martins FM, Benediktsson JA (2014) Multilevel image segmentation based on fractional-order darwinian particle swarm optimization. IEEE Trans Geosci Remote Sens 52(5):2382–2394
4. Mughees A, Tao L (2016) Efficient deep auto-encoder learning for the classification of hyperspectral images. In: 2016 international conference on virtual reality and visualization (ICVRV), pp 44–51. IEEE
5. Ghamisi P, Benediktsson JA, Ulfarsson MO (2013) The spectral-spatial classification of hyperspectral images based on hidden markov random field and its expectation-maximization. In: 2013 IEEE international geoscience and remote sensing symposium (IGARSS), pp 1107–1110. IEEE
6. Willett RM, Duarte MF, Davenport MA, Baraniuk RG (2014) Sparsity and structure in hyperspectral imaging: sensing, reconstruction, and target detection. IEEE Signal Process Mag 31(1):116–126
7. Van Der Meer F (2004) Analysis of spectral absorption features in hyperspectral imagery. Int J Appl Earth Obs Geoinf 5(1):55–68
8. Lacar F, Lewis M, Grierson I (2001) Use of hyperspectral imagery for mapping grape varieties in the Barossa Valley, South Australia. In: IEEE 2001 international geoscience and remote sensing symposium, 2001. IGARSS'01, vol 6, pp 2875–2877, IEEE
9. Malthus TJ, Mumby PJ (2003) Remote sensing of the coastal zone: an overview and priorities for future research
10. Yuen PW, Richardson M (2010) An introduction to hyperspectral imaging and its application for security, surveillance and target acquisition. Imaging Sci J 58(5):241–253
11. Kuybeda O, Malah D, Barzohar M (2007) Rank estimation and redundancy reduction of high-dimensional noisy signals with preservation of rare vectors. IEEE Trans Signal Process 55(12):5579–5592
12. Jia X, Kuo B-C, Crawford MM (2013) Feature mining for hyperspectral image classification. Proc IEEE 101(3):676–697
13. Chen Y, Lin Z, Zhao X, Wang G, Gu Y (2014) Deep learning-based classification of hyperspectral data. IEEE J Sel Top Appl Earth Obs Remote Sens 7(6):2094–2107
14. Hughes G (1968) On the mean accuracy of statistical pattern recognizers. IEEE Trans Inf Theory 14(1):55–63
15. Krizhevsky A, Sutskever I, Hinton GE (2012) Imagenet classification with deep convolutional neural networks. In: Advances in neural information processing systems, pp 1097–1105

16. Yuan Z, Lu Y, Xue Y (2016) Droiddetector: android malware characterization and detection using deep learning. Tsinghua Sci Technol 21(1):114–123
17. Gao Y, Dai Q (2014) Efficient view-based 3-d object retrieval via hypergraph learning. Tsinghua Sci Technol 19(3):250–256
18. Chen X, Huang H (2011) Immune feedforward neural network for fault detection. Tsinghua Sci Technol 16(3):272–277
19. Fauvel M, Tarabalka Y, Benediktsson JA, Chanussot J, Tilton JC (2013) Advances in spectral-spatial classification of hyperspectral images. Proc IEEE 101(3):652–675
20. Zhao W, Du S (2016) Spectral-spatial feature extraction for hyperspectral image classification: a dimension reduction and deep learning approach. IEEE Trans Geosci Remote Sens 54(8):4544–4554
21. Zhao W, Du S (2016) Learning multiscale and deep representations for classifying remotely sensed imagery. ISPRS J Photogram Remote Sens 113:155–165
22. Mughees A, Tao L (2017) Hyperspectral image classification via shape-adaptive deep learning. In: 2017 IEEE international conference on image processing (ICIP)
23. Mughees A, Tao L (2017) Hyper-voxel based deep learning for hyperspectral image classification. In: 2017 IEEE international conference on image processing (ICIP)
24. Chen Y, Zhao X, Jia X (2015) Spectral-spatial classification of hyperspectral data based on deep belief network. IEEE J Sel Top Appl Earth Obs Remote Sens 8(6):2381–2392
25. Li J, Bioucas-Dias JM, Plaza A (2013) Spectral-spatial classification of hyperspectral data using loopy belief propagation and active learning. IEEE Trans Geosci Remote Sens 51(2):844–856
26. Liu J, Wu Z, Wei Z, Xiao L, Sun L (2013) Spatial-spectral kernel sparse representation for hyperspectral image classification. IEEE J Sel Top Appl Earth Obs Remote Sens 6(6):2462–2471
27. Foody GM, Mathur A (2004) A relative evaluation of multiclass image classification by support vector machines. IEEE Trans Geosci Remote Sens 42(6):1335–1343
28. Mughees A, Chen X, Tao L (2016) Unsupervised hyperspectral image segmentation: merging spectral and spatial information in boundary adjustment. In: 2016 55th annual conference of the society of instrument and control engineers of Japan (SICE), pp 1466–1471. IEEE
29. Ambikapathi A, Chan T-H, Lin C-H, Chi C-Y (2012) Convex geometry based outlier-insensitive estimation of number of endmembers in hyperspectral images. Signal 1:1–20
30. Hinton GE (2002) Training products of experts by minimizing contrastive divergence. Neural Comput 14(8):1771–1800
31. Jia X, Richards JA (1999) Segmented principal components transformation for efficient hyperspectral remote-sensing image display and classification. IEEE Trans Geosci Remote Sens 37(1):538–542
32. Zabalza J, Ren J, Yang M, Zhang Y, Wang J, Marshall S, Han J (2014) Novel folded-PCA for improved feature extraction and data reduction with hyperspectral imaging and SAR in remote sensing. ISPRS J Photogram Remote Sens 93:112–122
33. Melgani F, Bruzzone L (2004) Classification of hyperspectral remote sensing images with support vector machines. IEEE Trans Geosci Remote Sens 42(8):1778–1790
34. Mou L, Ghamisi P, Zhu XX (2017) Deep recurrent neural networks for hyperspectral image classification. IEEE Trans Geosci Remote Sens 55(7):3639–3655
35. Hu W, Huang Y, Wei L, Zhang F, Li H (2015) Deep convolutional neural networks for hyperspectral image classification. J Sens 2015
36. Jimenez LO, Landgrebe DA (1999) Hyperspectral data analysis and supervised feature reduction via projection pursuit. IEEE Trans Geosci Remote Sens 37(6):2653–2667
37. Chang C-I, Du Q, Sun T-L, Althouse ML (1999) A joint band prioritization and band-decorrelation approach to band selection for hyperspectral image classification. IEEE Trans Geosci Remote Sen 37(6):2631–2641
38. Lin Z, Chen Y, Zhao X, Wang G (2013) Spectral-spatial classification of hyperspectral image using autoencoders. In: 2013 9th international conference on information, communications and signal processing (ICICS), pp 1–5. IEEE

39. Plaza A, Plaza J, Martin G (2009) Incorporation of spatial constraints into spectral mixture analysis of remotely sensed hyperspectral data. In: IEEE international workshop on machine learning for signal processing, 2009. MLSP 2009, pp 1–6. IEEE
40. Qian Y, Ye M (2013) Hyperspectral imagery restoration using nonlocal spectral-spatial structured sparse representation with noise estimation. IEEE J Sel Top Appl Earth Obs Remote Sens 6(2):499–515
41. Mughees A, Chen X, Du R, Tao L (2016) Ab3c: adaptive boundary-based band-categorization of hyperspectral images. J Appl Remote Sens 10(4):046009–046009
42. Derin H, Kelly PA (1989) Discrete-index Markov-type random processes. Proc IEEE 77(10):1485–1510
43. Ghamisi P, Benediktsson JA, Ulfarsson MO (2014) Spectral-spatial classification of hyperspectral images based on hidden Markov random fields. IEEE Trans Geosci Remote Sens 52(5):2565–2574
44. Bengio Y, LeCun Y et al (2007) Scaling learning algorithms towards AI. Large-Scale Kernel Mach 34(5):1–41
45. Camps-Valls G, Bruzzone L (2009) Kernel methods for remote sensing data analysis. Wiley
46. López-Fandiño J, Quesada-Barriuso P, Heras DB, Argüello F (2015) Efficient elm-based techniques for the classification of hyperspectral remote sensing images on commodity GPUs. IEEE J Sel Top Appl Earth Obs Remote Sens 8(6):2884–2893
47. Zhang Y, Brady M, Smith S (2001) Segmentation of brain MR images through a hidden Markov random field model and the expectation-maximization algorithm. IEEE Trans Med Imaging 20(1):45–57
48. Breiman L, Friedman J, Olshen R, Stone C (1965) Isodata, a novel method of data analysis and classification. Tech. rep., Technical report, Stanford Research Institute
49. Geman S, Geman D (1984) Stochastic relaxation, Gibbs distributions, and the Bayesian restoration of images. IEEE Trans Pattern Anal Mach Intell 6:721–741
50. Vincent P, Larochelle H, Lajoie I, Bengio Y, Manzagol P-A (2010) Stacked denoising autoencoders: learning useful representations in a deep network with a local denoising criterion. J Mach Learn Res 11(Dec):3371–3408
51. Tarabalka Y, Chanussot J, Benediktsson JA (2010) Segmentation and classification of hyperspectral images using watershed transformation. Pattern Recognit 43(7):2367–2379
52. Tropp JA, Gilbert AC (2007) Signal recovery from random measurements via orthogonal matching pursuit. IEEE Trans Inf Theory 53(12):4655–4666
53. Li J, Bioucas-Dias JM, Plaza A (2011) Hyperspectral image segmentation using a new Bayesian approach with active learning. IEEE Trans Geosci Remote Sens 49(10):3947–3960
54. Camps-Valls G, Tuia D, Bruzzone L, Benediktsson JA (2013) Advances in hyperspectral image classification: earth monitoring with statistical learning methods. IEEE Signal Process Mag 31(1):45–54
55. Zhong P, Wang R (2010) Learning conditional random fields for classification of hyperspectral images. IEEE Trans Image Process 19(7):1890–1907
56. Ratle F, Camps-Valls G, Weston J (2010) Semisupervised neural networks for efficient hyperspectral image classification. IEEE Trans Geosci Remote Sens 48(5):2271–2282
57. Gualtieri J, Chettri S (2000) Support vector machines for classification of hyperspectral data. In: IEEE 2000 international geoscience and remote sensing symposium, 2000. Proceedings. IGARSS 2000, vol 2, pp 813–815. IEEE
58. Rodarmel C, Shan J (2002) Principal component analysis for hyperspectral image classification. Surv Land Inf Sci 62(2):115
59. Wang J, Chang C-I (2006) Independent component analysis-based dimensionality reduction with applications in hyperspectral image analysis. IEEE Trans Geosci Remote Sens 44(6):1586–1600
60. Bengio Y, Courville A, Vincent P (2013) Representation learning: a review and new perspectives. IEEE Trans Pattern Anal Mach Intell 35(8):1798–1828
61. Ghamisi P, Benediktsson JA, Sveinsson JR (2014) Automatic spectral-spatial classification framework based on attribute profiles and supervised feature extraction. IEEE Trans Geosci Remote Sens 52(9):5771–5782

62. Ghamisi P, Benediktsson JA, Cavallaro G, Plaza A (2014) Automatic framework for spectral-spatial classification based on supervised feature extraction and morphological attribute profiles. IEEE J Sel Top Appl Earth Obs Remote Sens 7(6):2147–2160
63. Li J, Zhang H, Zhang L (2014) Supervised segmentation of very high resolution images by the use of extended morphological attribute profiles and a sparse transform. IEEE Geosci Remote Sens Lett 11(8):1409–1413
64. Li J, Marpu PR, Plaza A, Bioucas-Dias JM, Benediktsson JA (2013) Generalized composite kernel framework for hyperspectral image classification. IEEE Trans Geosci Remote Sens 51(9):4816–4829
65. Zhou Y, Peng J, Chen CP (2015) Extreme learning machine with composite kernels for hyperspectral image classification. IEEE J Sel Top Appl Earth Obs Remote Sens 8(6):2351–2360
66. Chen Y, Nasrabadi NM, Tran TD (2013) Hyperspectral image classification via kernel sparse representation. IEEE Trans Geosci Remote Sens 51(1):217–231
67. Tarabalka Y (2010) Classification of hyperspectral data using spectral-spatial approaches. PhD thesis, Institut National Polytechnique de Grenoble-INPG
68. Hinton GE, Salakhutdinov RR (2006) Reducing the dimensionality of data with neural networks. Science 313(5786):504–507
69. Yu D, Deng L, Wang S (2009) Learning in the deep-structured conditional random fields. In: Proceedings of NIPS workshop, pp 1–8
70. Mohamed A-R, Dahl G, Hinton G (2009) Deep belief networks for phone recognition. In: Nips workshop on deep learning for speech recognition and related applications, vol 1, p 39
71. Zhang L, Zhang L, Du B (2016) Deep learning for remote sensing data: a technical tutorial on the state of the art. IEEE Geosci Remote Sens Mag 4(2):22–40
72. Hinton GE, Osindero S, Teh Y-W (2006) A fast learning algorithm for deep belief nets. Neural Comput 18(7):1527–1554
73. Zhu Z, Woodcock CE, Rogan J, Kellndorfer J (2012) Assessment of spectral, polarimetric, temporal, and spatial dimensions for urban and peri-urban land cover classification using Landsat and SAR data. Remote Sens Environ 117:72–82
74. Lam L, Suen S (1997) Application of majority voting to pattern recognition: an analysis of its behavior and performance. IEEE Trans Syst Man Cybern Part A Syst Hum 27(5):553–568

Chapter 7
Sparse-Based Hyperspectral Data Classification

In this section, we restate the sparsest solution problem using a geometric interpretation. Finding the sparsest solution is strictly equivalent to the l_0-norm problem in Eq. 7.1. Unfortunately, this l_0-*minimization* problem is computationally intensive, so we will prove that the following l_1-*minimization* approach in Eq. 7.2 is a good approximation to it. Let's first define a unit polytope $\mathcal{P}_0$ as

$$\mathcal{P}_0 = \{\boldsymbol{\alpha} \ : \ ||\boldsymbol{\alpha}||_1 \leq 1\} \ \subset \ \mathbb{R}^n$$

By doing a linear transformation A, we obtain a new polytope $A(\gamma\mathcal{P}_0)$:

$$A(\gamma\mathcal{P}_0) = \{\boldsymbol{\alpha} \ : \ ||A\boldsymbol{\alpha}||_1 \leq \gamma\} \ \subset \ \mathbb{R}^m$$

In this sense, the l_1-*minimization* problem is equivalent to that when we scale γ from 0, to find the first γ^* such that $\boldsymbol{y} \in A(\gamma^*\mathcal{P}_0)$.

Note that this linear transformation has the following properties:

- Boundary Reserve Property: the nodes of the polytope $\mathcal{P}_0$ are still facets of $A(\mathcal{P}_0)$.
- Scaling Property: if we scale $\mathcal{P}_0$, the image under multiplication of A is also scaled by the same amount.

These two properties directly derive the following conclusion: l_1-*minimization* recovers the sparsest solution $\boldsymbol{\alpha}_0$ if $A(\boldsymbol{\alpha}_0/||\boldsymbol{\alpha}_0||_1)$ is a node of $A(\mathcal{P}_0)$, i.e., $\boldsymbol{\alpha}_0$ has only one non-zero entry.

This claim could be generalized. Let's defined $A(\mathcal{P})$ as t-neighborly if A maps all t-dimensional facets of $\mathcal{P}_0$ to facets. If so, we can conclude that

ℓ_1 correctly finds the sparsest solution with no larger than t non-zero entries if and only if $A(\mathcal{P}_0)$ is t-neighborly.

Note that this t is supposed to be quite high: even polytope $\mathcal{P}$ given by random coefficients (e.g., uniform and Gaussian) are highly neighborly.

L. Tao and A. Mughees, *Deep Learning for Hyperspectral Image Analysis and Classification*, Engineering Applications of Computational Methods 5,
https://doi.org/10.1007/978-981-33-4420-4_7

The classification of high-dimensional data is a major problem in pattern recognition. If the ratio of the given training data to the dimensional of data is limited, it becomes increasingly difficult to estimate the parameters, because the values of parameters become unreliable. As a result, the classification accuracies of the algorithms decline. In remote sensing, it is difficult to have a large number of labeled samples in general, as it involves going on ground, and assigning labels to the pixels or matching the spectra of the pixels to the spectral libraries, which needs an experienced analyst. As the new sensors have emerged in remote sensing, i.e., hyperspectral sensors, the problem has become more pronounced as the sensors capture the light across the electromagnetic spectrum in a large number of bands, and therefore produce high-dimensional data. These sensors, therefore, bring exceptional information for the classification of numerous materials on the ground, but at the same time it remains difficult and extremely expensive to assign labels to these samples. It is highly desirable to have a classifier which can work in high-dimensional space, and for a few labeled samples, because it would reduce the need to have a large number of labeled samples, and ultimately would reduce the cost and effort required for the classification of hyperspectral data. In this chapter, new classification approaches are proposed for the classification of hyperspectral data for a few labeled samples in high-dimensional space.

The grouping of high-dimensional information with not many marked examples is a significant test which is hard to meet except if some extraordinary attributes of the information can be exploited. In remote sensing areas, the issue is especially genuine due to the trouble and cost factors engaged with the task of labeling the high-dimensional data. This chapter exploits certain unique properties of hyperspectral information and proposes a ℓ^1***-minimization***-based sparse depiction classification way to deal with this constraint of very less number of training samples in comparison with the amount of spectral channels available in the hyperspectral information characterization. We accept that the information inside each hyperspectral channel information class lies in an exceptionally low- dimensional subspace. In contrast to conventional managed techniques, the proposed strategy doesn't have separate preparing and testing stages and, along these lines, doesn't need to bother with a preparation system for model creation.

Further, to demonstrate the sparsity of hyperspectral information and handle the computational escalation and time request of universally useful linear programming (LP) solvers, we propose a Homotopy-based sparsity order approach, which works proficiently when information is exceptionally inadequate.

The methodology isn't just time proficient. However, it additionally delivers results, which are tantamount to the conventional techniques. Extensive experiments on four real hyperspectral data sets show that hyperspectral data is highly sparse and, in most of the cases, the proposed approaches offer higher classification accuracy than state-of-the-art methods. Broad investigations on four genuine hyperspectral informational collections demonstrate that hyper otherworldly information is profoundly meager in nature, and the proposed approaches are strong across various databases, offer more accuracy, precision, and are more proficient than available cutting-edge techniques.

7.1 Introduction

In the course of the most recent decade, hyperspectral sensors have developed as a significant leap forward in remote sensing for giving important and rich data. These sensors acquire the brilliance of materials in an extremely enormous number of continuous channels and, accordingly, give magnificent data that contains information of different materials. Because of the rich data acquired by the sensors, hyperspectral classification has been applied in numerous regions.

In remote sensing, hundreds and thousands of marked samples are typically required to get great order results which are usually almost impossible to get and because of the extensive number of spectral data, the Hughes phenomenon occurs. Sadly, the availability of an adequate number of sampled tests is a general issue in computer vision and image processing, and this is a progressively serious issue in remote sensing since it is incredibly troublesome and costly to recognize and mark the training samples, and once in a while it isn't even plausible. Currently, in this chapter, we address a significant issue of utilizing less-sampled hyperspectral data, i.e., utilizing very less available training examples in high- dimensional space. Regardless of the enormous measure of research that has been accomplished for the classification of spectral data, the characterization of high-dimensional hyperspectral information, utilizing limited training samples, is yet an open research zone [1, 2].

For hyperspectral information classification, supervised strategies work rather well [3]. Artificial neural systems have been proposed, however, to get great classification results; enormous quantities of labeled training data are expected to get familiar with the model, and a fitting dimensional reduction procedure is required too. Wright et al. [4] proposed the sparse-based grouping approach for face recognition, which needs an enormous number of training samples. Others [5, 6] have utilized meager-based directed arrangement techniques for hyperspectral information characterization, however, those need a great deal of labeled data to function admirably.

In supervised techniques [7, 8], support vector machine (SVM) is a cutting-edge approach, which functions admirably by keeping itself within the limits of constrained training data and in the presence of high-dimensional information in the form of hyperspectral image. Notwithstanding, SVM-based strategies are delicate to model choice. It would be ideal if you allude to Section II for additional subtleties. Another technique that has of late been acquainted which handles the issue of less training data with some achievement is the dynamic subspace strategy [9], where a different classifier framework is presented for the characterization of hyperspectral information utilizing limited training data.

Even though numerous classification algorithms are accessible in writing, hyperspectral information classification by utilizing a few labeled samples in high-dimensional space is as yet a difficult task.

Another customary method to take care of the issue of scarce training samples for high-dimensional information classification is by reducing the dimensional through feature extraction or feature selection. The significant disadvantage of dimensional reduction strategies is that these work in explicit circumstances, and no broad generic procedure is accessible that suits all hyperspectral information. It needs a considerable amount of time, earlier information, and experimentation to look for a fitting dimensional reduction method.

In recent times, sparse depiction from over-complete word references has been the focal point of consideration in a computer vision network [10]. Sparsity is a powerful model that looks to depict information as a direct sequence of a fewer number of arrangements, which are found out from the information itself. To get an extraordinary depiction, symmetrical changes like the discrete cosine change and discrete wavelet change have been utilized in a signal preparing network for quite a while [11]. Late advancement right now demonstrated that sparse depiction utilizing over-complete word references can be comprehended through straight programming[12].

In the early stages of the second decade of the twenty-first century, sparse depiction was successfully experimented in data classification scenarios. So we can see the implementation of this concept in many computer vision areas ranging from face identification [4, 13], speech identification [14], and signal identification [15]. Similarly, the research in compressed sensing at that time [4, 7, 16–19] concluded after a detailed experimentation that if the potential result is scarce enough, then at that stage ℓ^1-minimization acquires a similar result as ℓ^0-minimization.

The issue of high dimensionality and scarce training samples is very difficult to defeat as a rule, except if the information contains some unique properties. In this chapter, we propose a ℓ^1***-minimizationn***-based sparse data classification technique to exploit extraordinary properties of hyperspectral information. The methodology utilizes a few available samples for high-dimensional data where a scarce depiction of each HSI image is evaluated by utilizing a word reference framed by the available training data of the considerable number of classes. Note that the meager-based methodology is definitely not a general arrangement approach, and its exhibition relies on the qualities of information. Our methodology is dependent on the expectation that training data examples, which have a place with a specific class around, lie in a low-dimensional subspace. Additionally, any test can be determined as the straight blend of fewer dictionary items. Consequently, in view of this expectation, we construe that to get great results by utilizing fewer training data labels, we can utilize the ℓ^1-minimization-focused sparse methodology because it will provide you the sparsest arrangement as far as word reference is concerned, i.e., right now, training data.

Authors in [5, 20, 21] presented a scarce- based representation for HSI classification. However, in contrast to their methodology, we are proposing the sparse-based classification framework for HSI information utilizing fewer available training samples in high-dimensional space. Along these lines, we are demonstrating the over-determined, i.e., $d > n$ situation as the under-determined issue, i.e., $d < n$ which additionally models the noise too.

It is accounted for in [22] that profoundly precise classification results for hyperspectral information grouping can be acquired by learning word references for each class when information is remade utilizing an exceptionally little level of training samples. This chapter focuses on Our proposed approach which depends on world-wide sparse presentation rather than nearby per class presentation, which contains the training samples of all classes in a single word reference.

The developed technique has the benefit of better performance in the classification of HSI because the scarce method possesses discrimination characteristics which have the ability to choose the fewest parameters from the given word reference (that means it possesses training samples of given classes) that effectively presents the training data. This developed algorithmic framework has even more advantage because it best fits the requirements of HSI data. Experimental results have shown very encouraging results for classification. Moreover, in contrast to the typical methods that requires dimensionality reduction of HSI and independent training and testing stages which are required to generate a model which is utilized afterwards for generating results, the developed methods don't require anything of this sort.

Since the universally useful straight programming (LP) solvers, i.e., inside point techniques and simplex, and so forth., are computationally expensive ($O(n^3)$) and tedious, they are less reasonable for quick processing. We, in this manner, further propose a quick scarce presentation utilizing ℓ^1-minimization dependent on *Homotopy* [23–25] for the classification of hyperspectral information. *Homotopy* calculation offers the preferred solution that if the hidden arrangement has k non-zeros, it accomplishes that arrangement in k iterative advances, i.e., the$k$-step property. Putting it in another way, *Homotopy* calculation works productively just when the arrangement is profoundly sparse. It will be demonstrated that *Homotopy*-rooted sparse presentation is exceptionally scarce for HSI information and a lot quicker than the generic LP solutions, while the precision is as yet practically identical to the universally useful LP solutions.

Broad tests on four genuine hyperspectral informational collections show that hyperspectral information are profoundly sparse, and, in the vast majority of the cases, the proposed approach offers higher classification performance than in the existing class techniques, i.e., SVM, semi-regulated, unique subspace, nearest neighbor, and nearest subspace strategies, even in the instances of fewer training samples. What's more, we additionally performed time correlations of the *Homotopy*-based methodology with general LP solvers, which obviously show that the methodology is a lot quicker.

The remaining chapter is formulated along these lines. Section 7.2 discusses the related techniques. Section 7.3 presents the recommended technique. Section 7.4 discusses the experiments and their results. Section 7.5.1 presents the sparsity of evaluated solver. Lastly, Sect. 7.6 provides conclusion of the chapter.

7.2 Related Methods

This part of the chapter presents and briefly describes the related techniques including nearest subspace and nearest neighbor strategies. These are identified with the ℓ^1-based scarce strategy. Three different techniques, i.e., dynamic subspace, SVM, and semi-administered strategies, are likewise presented. These are cutting-edge strategies proposed for hyperspectral information classification utilizing less training samples. *Nearest Subspace Method (NS) and Nearest Neighbor Method (NN)*: NS [26] presumes that the samples of a specific class are contained in a subspace, and those samples cover the subspace, i.e., the samples go about as the premise of the subspace. Any new test vector can be taken as the direct blend of the premise vector, i.e., or equivalently the solution can be described as

$$y = \Sigma_{j=1}^{n} \alpha . a_j + \epsilon$$

or equivalently

$$y = B\alpha + \epsilon$$

The solution can be presented as

$$\hat{\alpha} = argmin_{\alpha} \left\| y - B\alpha \right\|_2^2$$

It very well may be designed as relevant to the ℓ^1-minimization scarce technique as the two strategies are relied upon to pick a similar class when perfect conditions exist for the information. Hence, under the conditions, the determined sparse presentation has the non-zero passages for an individual class. NS expects the scarce-land architecture [27], i.e., just the items in the scarce presentation compared to its group are non-zero. Be that as it may, truly, such perfect conditions never exist .

The Neural Network technique associates the label of the closest classified sample to the test sample.

$$f(B, x) = min \left\| x - b \right\|_2^2$$

It can be demonstrated by exploring the similar techniquess that nearest neighbor and ℓ^1 based classifier are expected to choose the same class when almost 1-sparse spare representation exists, i.e., roughly with a single non-zero item. The NN strategy makes the presumption that the determined scarce presentation is 1-scarce. In any case, again, such perfect conditions never exist in genuine situations.

Support Vector Machines(SVM): SVM [7, 8] is a famous classification technique based on supervised learning that performs the prediction by following the equation:

$$y(a, w) = \Sigma_{j=1}^{n} w_j L(a, a_j) + w_0$$

Here, w_j presents the architecture weights and $L(.,.)$ presents a kernel which depicts a premise work for each marked sample. In SVM, the objective capacity not just attempts to limit the error for a marked set but also attempts to expand the edge between two classes, which are characterized in the element space by a part L. As a consequence of this procedure, a *sparse architecture* is received that is subject to a subset of kernel equations that are connected with the examples which lie on either side of the edge. The relevant examples are known as support vectors. One can assume that frugality [4] is likewise the main consideration in the SVMs. SVMs [28] have been known to have magnificent execution in the hyperspectral information classification utilizing less marked examples. Nonetheless, the selection of a specific kernel is really critical for its performance that implies that it is kernel dependent. Moreover, selecting the optimal parameters for the execution is also very important.

Dynamic Subspace Method (DSM): DSM is another proposed technique by [9] for HSI classification in the presence of fewer training data. It depends on the irregular subspace technique dynamic subspace method (DSM) and utilizes an outfit of classifiers for the HSI classification. It utilizes two appropriations. The first is utilized to demonstrate the likelihood of band choice, i.e., it depends on the rule that the valuable channels convey higher probabilities than the non-helpful channels in HSI. The subsequent dispersion is utilized for picking subspace dimensionality, i.e., the quantity of channels. At that point, for every classifier in the troupe, the new subspace dimensionality and the diminished measurement set are drawn from the two circulations. An ultimate choice depends on majority voting. It has demonstrated great outcomes for the hyperspectral data classification in the presence of fewer samples. In DSM, the exactness of results is vigorously reliant on the estimation of the distributions.

Semi-Supervised Graph-Based Method for Hyperspectral Classification: The Semi-administered Graph-Based Method for Hyperspectral Classification strategy [28] has been developed to handle the exceptional attributes of hyperspectral information, i.e., high dimensionality and availability of less training data. It utilizes the kernel rooted technique to deal with high dimensionality. It is a semi-supervised strategy to exploit unlabeled training information. Lastly, it utilizes composite kernels to compensate spatial data. The strategy analyzes its results with the very famous SVM technique in the presence of fewer samples.

7.3 Proposed Approach

The sparse presentation which depends on the sparse-land architecture [27] utilizes training examples itself to figure out the depiction of test data [4]. In this section of the chapter, we present a sparse presentation for hyperspectral image classification in the presence of fewer training data.

7.3.1 *Sparse Representation for Hyperspectral Data Using a Few Labeled Samples*

In the sparse depiction, the essential concern is to have an arrangement of linear mathematical equations wherein the volume of unknown parameters is greater than the volume of equations. In other words, it is a scenario of under-resolved parameters, i.e., $d < n$. There could be numerous resolved possibilities for such scenario, however, the most attractive arrangement is the one having the least conceivable non-zeros [12]. We present a sparse rooted technique for hyperspectral image classification for solving the significant problem of high dimensionality and availability of less training data. In particular, the trouble is very difficult to conquer except if some interesting attributes exist in the data.

In general, the framework based on sparse representation is not particularly developed for classification. Moreover, the accuracy of the approach is dependent on the scarcity of the data. There is a high possibility that the sparse focused approach delivers the best results if the given data is highly sparse. Here, our technique is exploring the unique characteristics of the HSI. Every class of HSI consists of samples that are identical in nature. This is due to the fact that the same material is reflected on the earth. Therefore, HSI data doesn't possess significant variation in the light of training data. It therefore can be deduced that fewer labeled samples are sufficient to compensate all the divergence in the given testing data. Hence, we conclude that in reality, HSI spectral–spatial information is rich in sparsity, therefore, the sparsity-based technique can be effectively applied for efficient HSI classification. We have taken one supposition into account here, where each class corresponds to a low-dimensional subspace. With this supposition, we can further assume that the test sample of HSI possesses high sparsity depiction corresponding to dictionary items. We have performed extensive experiments to prove this supposition.

Let's suppose we have C distinctive classes and each label of every class c_j shows particular material on earth. As indicated by the suspicion, the dissemination of tests of each class frames a low-dimensional linear subspace, which can be spread over by fewer basis vectors, collecting all basis items into a structure known as a dictionary. The test samples, which are m-dimensional vectors, require to be classified into c_j classes. This multi-dimensional vector actually represents the reflected value that is calculated by the hyperspectral remote sensor along various wavelength ranges. Fewer training samples (a_j, c_j) are available for every class. Here, a_j is the available training data and c_j is the specified label. The scarce presentation of a provided training item x is evaluated as $B\alpha = x$. Here, α is describing the sparse depiction of vector x. Now, to evaluate the sparse depiction of test data, given labeled data is arranged in a 2D matrix $B = [b_1, b_2, \ldots, b_n]$.

In order to identify the sparse depiction of each test item x, vector α requires to be calculated for $x = B\alpha$. Entries which are not zero in the subject vector can be utilized to reproduce the test item, which can further be utilized for classifying the HSI data. In order to identify the vector comprising scarce values, this becomes an optimization scenario:

$$\alpha_0 = \text{argmin} \parallel \alpha \parallel_0 \text{ subject to } x = B\alpha \tag{7.1}$$

This could also be called ℓ^0-norm whose responsibility is to calculate the number of elements in the vector which are not zero. The optimized value of ℓ^0-norm is nondeterministic polynomial time (NP) hard. It is so difficult that its optimized values can not even be approximated as shown in [29]. Partially, the main cause of this is a non-convex nature of the ℓ^0 for the optimization scenario. However, recent advancements in compressed sensing [16, 17] have led us to solve or reduce this difficulty by receiving an estimated result if we reinstate ℓ^0-norm with ℓ^1-norm. A resultant new equation is a convex based formulation that can be resolved and optimized comparatively instantly through linear programming techniques:

$$\alpha_1 = \text{argmin} \parallel \alpha \parallel_1 \text{ subject to } x = B\alpha \tag{7.2}$$

Practically speaking, the received HSI pixel x is relatively always a mixer of some noise. There are many types of noises present, e.g., atmosphere and sensor-based noise. Therefore, the test item x cannot cover only by the linear integration of labeled data of the specified subject classes. Hence, we include the noise phrase η in the previous as

$$x = B\alpha + \eta \qquad \text{so as} \quad \parallel \eta \parallel_2 < \epsilon \tag{7.3}$$

Hence, the scarce resolution α can be constructed by utilizing ℓ^1-minimization as

$$\alpha_1 = \text{ argmin } \parallel \alpha \parallel_1 \text{ subject to } \parallel A\alpha - y \parallel_2 \ \leq \epsilon \tag{7.4}$$

Here, ℓ^1-norm could be formalized as

$$\parallel \alpha \parallel = \Sigma_i \mid \alpha_i \mid$$

Equation 7.4 is intimately associated with the below dissipated optimizational problem [30]:

$$\min_{x} \quad \frac{1}{2} \parallel x - B\alpha \parallel_2^2 + \lambda \parallel \alpha \parallel_1 \tag{7.5}$$

Here, λ acts as a regularizer which works as a equalizer between regeneration error resolution sparsity. If one inspects our proposed approach of hyperspectral image classification in the presence of limited training data and high dimensionality, closely, they could easily deduce that this solution is actually an over-determined case. That means there exist d and n such that $d > n$. However, to make our case under-determined, we exploited the compressed sensing technique in order to enhance the dimensionality of B by incorporating M. Hence, Eq. (7.2) can now be formalized as

$$x = [B \quad M] \begin{bmatrix} \alpha \\ \eta \end{bmatrix} = Ah \tag{7.6}$$

Here $A = [B \ M] \in R^{d\times(n+d)}$ and $h = [\alpha^T \ \eta^T]^T \in R^{n+d}$
Therefore, the scarce presentation can be evaluated through the below ℓ^1-norm scenario:

$$\min_h \parallel \text{h} \parallel_1 \quad \text{subject to } Ah = x \tag{7.7}$$

where this Eq. 7.6 contemplates two aims. First, it achieves the task of making our technique under-determined. Secondly, it also takes care of the noise that is present in the image. In the specified equation, $\eta \in R^d$ represents the error vector. This vector actually models the subject noise in channels of the HSI image. Subsequently, the error vector is integrated into the below objective function.

$$f_j(x) = \| x - B\hat{\alpha_j} - \eta \|_2 \quad j = 1, \ldots, m \tag{7.8}$$

This ℓ^1-minimization issue can be resolved through an internal point technique known as ℓ^1-magic [31]. The scarce presentation of each training data x is calculated by utilizing labeled data of every class. Estimation $B\hat{\alpha_j}$ of test vector x is constructed by utilizing the feature vector α_j. This vector contains exclusively the coefficients which are not zero and which are also associated with the indices of the core of a particular class. This estimated value for every class is deducted from the real test vector. The objective function is first applied and a minimized value is obtained for a particular class. That particular class label is attached to the test vector. Comprehensive experimentation has demonstrated the effectiveness of our technique in the presence of limited labeled training data. A complete framework of the proposed technique is presented in an algorithm.

Algorithm 1 Classification of Hyperspectral Data

Input: labeled samples,labels (x_i, c_i) $i = 1, \cdots, n$ and test vector y.

1: Normalize all the labeled samples x_i and test sample y.
2: Convert the case into underdetermined form using eq 3-5
3: Solve the ℓ^1-minimization problem.
4: Compute $f_i(y), i = 1, \cdots, n$.

Output: $\arg\min_i f_i(y)$.

7.3.2 Homotopy

Despite the fact that sparse presentation computation has been successfully decreased to the ℓ^1-norm optimization scenario, however, resolving it effectively is yet a very difficult and complex issue. Typically, detecting the sparse resolution has always been identified as a Combinatorics optimizing issue, although Eq. (7.2) can easily be resolved through linear programming (LP). Hence, productively it is more controllable in particular [24]. Nonetheless, the linear programming solution finally includes a resolution of nyn linear systems. It also desires many forms of the resolution, where each form costs an order of cubic, i.e., $O(n^3)$ flops [24]. The technique presented resolves ℓ_1-norm through LP by utilizing a generic linear programming resolver. That means a very simple and internal mark technique that results in the resolution of the entire nYn linear scheme in the order $O(n^3)$. Nonetheless, the speed of such optimizing solutions is not fast and this factor plays a significant role. For this reason, numerous other estimation resolution techniques have been presented: for example, quadrate identical pursuit technique [32] and varying direction technique [33]. In this scenario, we developed an effective sparse resolution, focused on a rapid and vigorous technique called *Homotopy*. This method is specially developed for the ℓ^1-norm optimizing scenario that is expected to contain a sparse resolution [34]. The promising capability of the subject design is dependent on the scarcity of the resolution. The *Homotopy* technique is formally developed by [24, 35]. It was developed especially to resolve the noise-based over-decisive ℓ^1-penalized least square problem (LSP). We have restated it here for hyperspectral image classification. Moreover, utilizing it to receive a robust ℓ^1-minimization focused on the resolution for classification. Additionally, it is also used to demonstrate the deeply sparse nature of the hyperspectral image.

Theoretically based on the sparseland assumption [27], i.e., an effective estimation of the real data can be efficiently generated through feature bank and atom collection by utilizing fewer atoms from the feature bank. This implies that *Homotopy* proves to be a very effective algorithm for our approach due to the fact that we have assumed that most of the data exists in lower dimensional subspace which is also proved by our comprehensive testing results for the said supposition. This algorithm *Homotopy* can be summarized as repetitively evaluating the resolution α as long as the predetermined precision-based condition is met which is focused on the original value $\alpha = \mathbf{0}$. It has also been demonstrated that resolving Eq. (7.2) is identical to resolving the following:

$$\min_{\alpha} f_{\lambda}(\alpha) = \min_{\alpha} \left\{ ||x - B\alpha||_2^2 + \lambda ||\alpha||_1 \right\} \tag{7.9}$$

Explicitly, the resolution of $f_{\lambda}(\alpha)$ coincides with the resolution of Eq. (7.2) [34], if $\lambda \to 0$. Hence, it is proved that the issue of equation (7.2) is decreased by resolving equation (7.9). In order to identify minima, it is essential that $\partial_{\alpha} f_{\lambda}(\alpha_{\lambda}) = 0$. The fractional disparity of $f_{\lambda}(\alpha)$ can be evaluated as

$$\partial_{\alpha} f_{\lambda}(\alpha_{\lambda}) = -B^{Q}(x - B\alpha_{\lambda}) + \lambda \mu(\alpha_{\lambda}) \tag{7.10}$$

Here, $\mu(\alpha) = \partial ||\alpha||_1 \in \mathbb{R}^n$, and the jth item of it is characterized as

$$\mu_j(\alpha) = \begin{cases} \text{sgn}(\alpha_j) & \alpha_j \neq 0 \\ \epsilon \in [-1, 1] & \alpha_j = 0 \end{cases}$$

For reducing and clarifying Eq. (7.10), one can characterize an index array $P = \{j : \alpha_{\lambda}(i) \neq 0\}$, i.e., items that are not zero, and enduring interrelation $d = B^{P}(x - B\alpha_{\lambda})$. Hence, along these lines, Eq. (7.10) can be rephrased as

$$e(M) = \lambda \cdot \text{sgn}(\alpha_\lambda(M)), |e(M^e)| \leq \lambda \tag{7.11}$$

On the gth phase, *Homotopy* calculates the new control f_p by resolving

$$B_M^P B_M f_g(M) = \text{sgn}(e_(M)) \tag{7.12}$$

Algorithm 2 *Homotopy* Algorithm

1: **initialize:** $\alpha \leftarrow 0, I \leftarrow \emptyset$
2: {we can choose other termination condition:}
3: **while** $\lambda > \lambda_0$ **do**
4: Compute $r_l^+, r_l^-, i^+, i^-, r_l$ by equation (3-12), (3-13), (3-14)
5: Compute d_l by equation (3-11)
6: Update I by appending i^+ or i^-
7: Update $\lambda \leftarrow \lambda - r_l$
8: Update $\alpha \leftarrow \alpha + r_l d_l$
9: **end while**

and consequently, items that are not present in M of f are assigned a zero value. Equation (7.11) is considered as an entire indexing situation of amending M when modifying α. It can easily be demonstrated and proved that two problems will drive to a contravention of Eq. (7.11); this happens if

$$s_P^+ = \min_{j \in M^e} \left\{ \frac{\lambda - e_P(j)}{1 - b_j^P w_P}, \frac{\lambda + e_P(j)}{1 + b^U w_P} \right\} \tag{7.13}$$

Here $w_P = B_M f_P(M)$, or

$$s_P^- = \min_{j \in M} \{ - \frac{y_P(j)}{f_P(j)} \} \tag{7.14}$$

In order to make things easier, define

$$s_P = \min\{s_P^+, s_P^-\} \tag{7.15}$$

Consequently, M can be modified by stacking index j^+ which makes Eq. (7.13) hold if $s_P^+ \leq s_P^-$; alternatively, we stack j^- to M. In parallel to this, one can modify α by utilizing equation (7.16):

$$\alpha_P = \alpha_{P-1} + s_P f_P \tag{7.16}$$

Lastly, one has to modify λ, i.e., $\lambda_P = \lambda_{P-1} - s_P$.

The completion situation of *Homotopy* can be organized in a couple of ways:

- Utilizing a predetermined short item λ_0, the algorithm stops if $\lambda_P \leq \lambda_0$;
- The algorithm stops if we control the threshold change of f in consecutive repetitions.

The step-wise detail of the proposed approach is presented in Algorithm 2. *Homotopy* rooted sparse presentation for training data is evaluated by utilizing Algorithm 2.

Algorithm 3 Sparse representation based classification using *Homotopy*

Input: labeled samples,labels (x_i, c_i) $i = 1, \cdots, n$ and test vector y.

1: Normalize all the labeled samples x_i and test sample y.

2: Solve the ℓ^1-minimization problem using *Homotopy* algorithm 2.

3: Compute $f_i(y), i = 1, \cdots, n$.

Output: $\arg\min_i f_i(y)$.

The complexity of the abovementioned algorithm is $O(u^3 + n)$. Here, u represents the quantity of items which are not zero in α. Therefore, we can observe that this process heavily decreases the complexity range from n^3 to u^3; provided that the quantity of items which are not zero is near to a fixed value, this implies *Homotopy* will be near to the direct pace. The time consumed for classification by utilizing *Homotopy* rises linearly if we raise the quantity of the training data, in contrast to the exponential rise in speed in repetition-based methods where the rise in time increases exponentially. The progress and cut-down in classification time by the proposed technique are not due to the trade-off between time and accuracy. The *Homotopy* technique is hypothetically factual and exact as being proved by equalizing Eq. (7.9). Moreover, the realistic accuracy is dependent on the stop criteria. One can adjust it by hand, i.e., it is not automatic to a value which has the same accuracy as internal spot techniques. Extensive experimentation demonstrates that the performance can easily be compared to the internal spot techniques.

7.3.3 Sparse Ensemble Framework

Integrating several classifiers for effective classification formally known as the ensemble classification technique is common in the research community for better performance. This method is based on the framework where prediction from every different classifier is received and the final decisive measure is taken depending on the output of all classifiers mostly based on majority voting. In the proposed technique, in order to classify each pixel, we have established an ensemble of separate sparse classifiers, named as the *Ensemble Classification Algorithm* (Algorithm 4). In the proposed technique, we contain a group of Q training data. Assume that S is a group of given training data. The technique consists of 3 comprehensive repetitions. Every repetition contains 7 sub-repetitions, i.e., each for every separate ensemble classifier. Each of the ensemble classifiers receives an irregular input from a given group of Q training data. In the initial comprehensive repetition, 7 separate classifiers generate their predicted result for the provided training data $x_j \in Q$. Moreover, the predicted result for each training data is saved as $d_{j1}, d_{j2}, \ldots, d_{j7}$. Subsequently, the majority voting criteria are employed after the initial comprehensive repetition is accomplished. That means voting is applied to the results received from the 7 scarce classifier. Moreover, the threshold β is characterized to calculate the major vote. For example, for a provided class label d, fulfilling $\{i : d_{ji} = d\}$ contains at minimum β votes, which implies that there exists a consensus, between the classifiers. Hence, we mark the class of x_j as d and discard x_j from the group Q. In contrast, if the consent is not achieved, we carry the data in Q. Therefore, following this technique, non-consensus data receives more weight.

In order to expand the training data and to induce variation in the data for the following repetitions, the below tests are applied to expand Q for each comprehensive repetition. For every class, d classified item can be detected with the least silt, i.e.,

$$x^{(d)} = \underset{x \notin S}{\operatorname{argmin}}\, g_j(x)$$

Here $g_j(x)$ is already characterized in Eq. (7.8). We modify Q

$$Q \leftarrow Q \bigcup_d x^{(d)}$$

Therefore, along these lines after each repetition, the training data, that is able to receive maximum votes and minimum leftover for a specific class, are included in the training data. This expanded training data is then made accessible for the indiscriminate excerpt of the training sample for subsequent repetitions. Each next repetition after the preceding repetition handles only that training data which is not in consent by each particular classifier in the preceding repetitions.

The below passage is narrated to demonstrate the authenticity of the proposed technique, particularly for a limited training data. As mentioned before, as a result of applying the method, we contain a collection of Q of training data. Every time, we select c of them as training data. Let's assume that the selection is indiscriminate such that accurate performance is roughly q every time. The classifier is run $(2L + 1)$ times to receive the maximum votes. 0/1 classification is presumed for the matter of simplification. Hence, current received accuracy q' can be written as

$$q' = \sum_{j=L+1}^{2L+1} \binom{2L+1}{L} q^j (1-q)^{2L+1-j} \tag{7.17}$$

Assume $s = \frac{q}{1-q}$, it can be assumed here that the real scarce can receive an efficiency of $\frac{1}{2}$ therefore $s \geq 1$. Currently, one can calculate the recent rate $s' = \frac{q'}{1-q'}$:

Table 7.1 Theoretical improvement of ensemble algorithm when $2k+1=7$

q	q'	Δq	q	q'	Δq
0.50	0.5000	0.0000	0.55	0.6083	0.0583
0.60	0.7102	0.1102	0.65	0.8002	0.1502
0.70	0.8740	0.1740	0.75	0.9294	0.1794
0.80	0.9667	0.1667	0.85	0.9879	0.1379
0.90	0.9973	0.0973	0.95	0.9998	0.0398

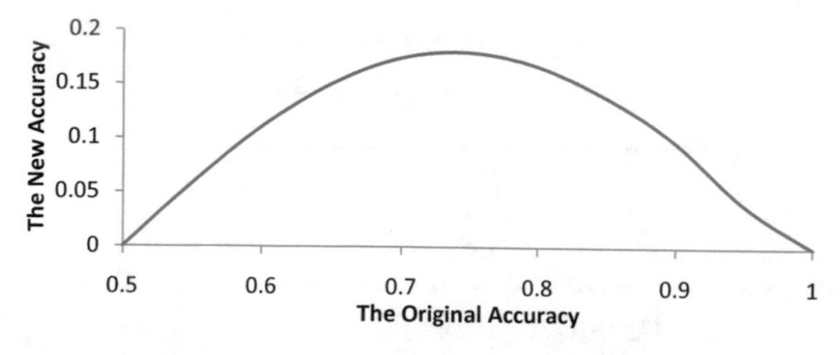

Fig. 7.1 Theoretical improvements

$$s' = \frac{n_0 r^{L+1} + n_1 s^{L+2} + \cdots + n_{L-1} s^{2L} + n_L s^{2L+1}}{n_0 s^L + n_1 s^{L-1} + \cdots + n_{L-1} s^1 + n_L s^0} \tag{7.18}$$

Here $n_j = \binom{2L+1}{L-i}$.

As $s \geq 1$, we can easily observe that $s' \geq s$, i.e., there are few developments. Table 7.1 depicts the theoretical bounds that one can accomplish by utilizing seven repetitions.

Figure 7.1 demonstrates the developments $\Delta q = q' - q$. One can observe that it attains the maximum if q is approximately 0.75. This clearly shows the reason behind the improvement in the efficiency of a few training data made by the ensemble technique. Assume, one utilizes 80% of the training data in every repetition, given that, the range $m =$, the range of every training data is 4, $q \approx 0.75$, the real development from q to q' over the peak. The whole procedure of the ensemble technique is illustrated in Fig. 7.2.

Algorithm : Ensemble classification

- Input set of training samples pool (x_j) $j = 1, \ldots, n$ and test vectors y_j.
- Normalize all the training samples x_i and test sample y.
- Randomly choose d training data subsets from the training data pool.
- Apply sparse classifiers using training subsets on each test sample and store the predictions of each classifier.
- Decide the class of each test sample on the basis of majority voting if there is consensus among the individual classifiers, otherwise mark the test sample as non-agreed.

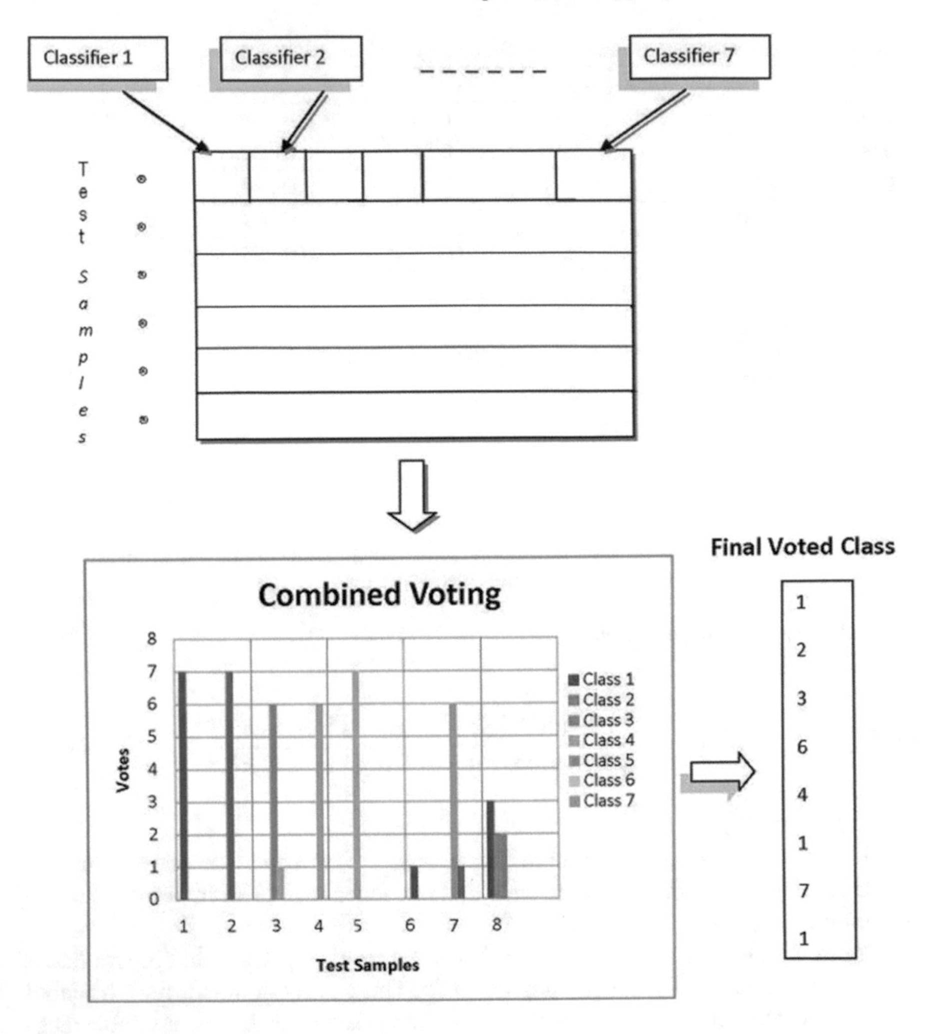

Fig. 7.2 Working of ensemble algorithm

- Add the newly classified samples, which have the smallest residuals for any particular class, in the training samples pool and repeat the whole process with only non-agreed test samples.
- Output: Final voted test samples labels.

Algorithm 4 Ensemble Framework

Input: Collections of training samples $\mathcal{P}$ and test vectors $\mathcal{T}$

1: **initialize:** Normalize all train and test vectors in $\mathcal{P} \cup \mathcal{T}$
2: **while** $\mathcal{T} \neq$ **and** number of iterations ≤ 3 **do**
3:     **for** $y \in \mathcal{T}$ **do**
4:         **for** each sparse classifier $1 \leq i \leq 7$ **do**
5:             randomly choose training set from $\mathcal{P}$
6:             classify y to c_i
7:         **end for**
8:         **if** $vote(c_1, \cdots, c_7) \geq \theta$ **then**
9:             $Class(y) \leftarrow major(c_1, \cdots, c_7)$
10:             $\mathcal{T} \leftarrow \mathcal{T} - \{y\}$
11:         **end if**
12:     **end for**
13:     $y^{(c)} = \arg\min_{y \notin \mathcal{T}} h_i(y)$
14:     $\mathcal{P} \leftarrow \mathcal{P} \bigcup_c y^{(c)}$
15: **end while**

Output: Class(y)

7.4 Experimental Results and Comparison

This part of the chapter explains in detail, the comprehensive experimentation that was conducted in order to assess and demonstrate the efficiency of the developed framework. The experimentation was performed on a Pentium-4 computer including Intel 2.4G Central Processing Unit with 2-Giga bytes of Random Access Memory running Windows XP. MATLAB R14 is utilized for scarce and *Homotopy* rooted implementation. We exploited 4 famous hyperspectral image-based datasets. These include AVIRIS sensor- based Indian Pine dataset, HYDICE sensor-based Washington DC Mall HSI image, Kennedy space center, and finally Salinas Valley HSI; these were utilized for experimentation and comparing with existing state-of-the-art methods. Every final result is based on the mean of 10 experiments which are conducted for that particular result. Moreover, Training data is randomly selected for each experiment. We do this in order to avoid any training data dependency for a particular result. The resulted performance of the proposed technique is great and strong for many abovementioned HSI images. The performance is compared with existing renowned techniques including support vector machine (SVM), semi-supervised techniques, dynamic subspace method, and neural network-based and NS techniques. SVM is one of the renowned techniques as its results are well established for the HSI classification. Comparison results for individual techniques are taken from the authors' own research papers: for example, findings of dynamic subspace technique from [9], and findings of SVM and SSM are captured from [28]. In other findings of SVM, the RBF kernel is explored, while 9-bed cross-validation is exploited for parameter choice. The parameter collection procedure can be studied thoroughly in [28]. The range of the selected kernels is in the limits of 10.3, ..., 103, while regularizer term ranges from 100, ..., 103. The rest of the parameters are created by utilizing an implementation code supplied by [28].

The following sub-parts of the chapter investigate the function of item collection in the scarce rooted HSI classification performance of HSI. After this, comprehensive experimentation on 4 hyperspectral images is conducted.

7.4.1 Effect of Parameter Selection on Classification Accuracy

The correctness of the classification of scarce rooted presentation depends on a couple of items. Those items include regularizer classification accuracy of a sparse-based representation is λ and fault resilience ϵ. The regularizer λ determines the compensation among scarcity of the resolution and the residual fault. The provided λ is big, and the resolution is going to be scarcer. However, the trade-off is that it escalates the residual fault because of which the real result is invisible. If λ is small, variation facts would be vanished. This is because of this reason that the result is going to be compressed. However, a small value decreases the residual fault.

Comprehensive and detailed experimentation was conducted to investigate the result of parameter choice on the correctness of the HSI result. In these experiments, correctness is measured at several parameter values λ and ϵ. That includes λ and ϵ, i.e., ϵ = e-01, e-02, e-03, e-04, e-05, and e-06, while the value of the regularizer is changed from 0.001 to 0.15 with incremental value of 0.01. The performance accuracy is depicted in Fig. 7.3. It can be observed from the figure that when the threshold is big, the performance decreases. The performance starts increasing when the threshold started decreasing. The performance is really small at ϵ e-01, e-02, i.e., the performance is approximately 76%. When ϵ is assigned e-03, the performance expands significantly. However, the performance doesn't increase linearly but there

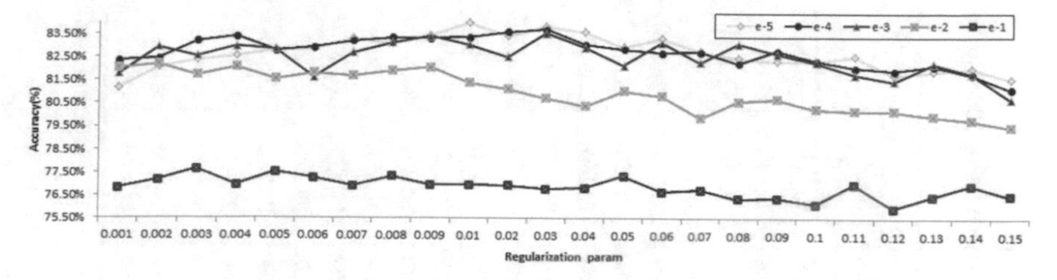

Fig. 7.3 Effect of parameters on classification accuracy

1	2	3	4	5	6	7	8	9
0.9662	0.9453	0.9935	0.9725	0.8708	0.9606	0.9574	0.9761	0.9457

Fig. 7.4 Energy covered by 5 PCs per class PCA for 9 classes of the AVIRIS Indian Pines dataset

is a variation in it. If ϵ is assigned e-04 and e-05, the performance stretches to the maximum. Afterwards, if the threshold is decreased further, it doesn't affect the performance. In the case of the the regularizer, the performance stretches to a maximum when the argument values are in the limit of 0.009 to 0.03. Afterwards, the performance begins to decline if the value of λ is additionally increased.

In order to get a further detailed investigation of the performance of every class of AVIRIS HSI which contains 9 distinct classes in total, relative to numerous argument groups of scarce rooted HSI classification, the performance for λ in limits of 0.006 to 0.06 and fault resilience at e-05, e-04, and e-05 are presented in Table 7.2. It can be clearly seen from the table that a maximum performance of 83.97% is acquired in the group 25, where λ is 0.05 and resilience is e-05. The confusion matrix of the same group is depicted in Table 7.3. This presents how much of the data is wrongly classified. For example, 81 data values of Corn-no-till are wrongly classified as Corn-min (Fig. 7.4).

7.4.2 *AVIRIS Hyperspectral Image*

The initial dataset that was utilized for the classification task is an HSI image that was captured through an AVIRIS sensor in the early 90s over the area of Indian Pine in Indiana. IT comprises 145 X 145 rows and columns and a spectral resolution of 220 bands. Bands that totally consist of noise (104–108, 150–163, 220) because of absorbing the water are discarded. Hence. as a result, 200 bands are explored for the subject task. This HSI image presents a very complex and demanding land classification problem, where the prime yields of the region (mostly corn and soybeans) were right into their initial development sequence, which contains only approximately five percent shade shelter. Differentiating between the main yields below these conditions is very challenging and complex (particularly, when a very narrow spatial

Table 7.2 Effect of error tolerances and regularization param on sparse-based classification accuracy for AVIRIS Hyperspectral Image using 100 labeled samples per class

Batch#	Err. Tol. (ϵ)	Reg. param (λ)	Classes									Acc (%)
			Corn-No-till	Corn-min	Grass/Pasture	Grass Trees	Hay-Windrowed	Soybeans-notil	Soybeans-min	Soybeans-Clean	Woods	
1	E-03	.006	75.71%	81.74%	89.17%	97.37%	100%	77.42%	66.51%	92.41%	99.33%	81.60%
2		.007	74.44%	81.47%	95.21%	98.92%	100%	79.15%	69.89%	90.47%	98.49%	82.66%
3		.008	78.86%	80.79%	94.46%	98.30%	100%	81.45%	68.20%	90.86%	99.16%	83.10%
4		.009	78.71%	84.47%	92.44%	98.92%	100%	77.30%	69.81%	89.49%	99.75%	83.37%
5		0.01	77.74%	81.34%	94.46%	99.23%	100%	84.45%	67.61%	87.55%	99.33%	83.01%
6		0.02	75.86%	80.52%	92.19%	99.69%	100%	83.06%	71.37%	89.69%	98.66%	83.52%
7		0.03	76.84%	84.47%	93.70%	100.00%	100%	83.29%	67.99%	86.38%	98.16%	82.91%
8		0.04	75.11%	78.07%	92.70%	98.92%	100%	77.53%	70.73%	86.58%	98.07%	82.13%
9		0.05	77.06%	84.33%	95.97%	99.69%	99.49%	82.49%	68.96%	85.21%	98.16%	83.11%
10		0.06	72.49%	80.11%	94.71%	99.07%	100%	83.87%	68.88%	88.13%	98.41%	82.27%
11	E-04	.006	77.75%	84.47%	92.44%	98.39%	99.85%	77.74%	68.72%	89.77%	99.51%	82.90%
12		.007	77.75%	82.07%	93.30%	99.29%	99.90%	78.85%	69.83%	90.16%	98.99%	83.18%
13		.008	78.38%	81.47%	94.06%	98.64%	99.74%	80.37%	69.29%	90.51%	99.51%	83.30%
14		.009	79.04%	81.61%	94.81%	98.86%	99.85%	79.40%	69.05%	90.97%	99.13%	83.28%
15		0.01	78.52%	81.69%	93.80%	99.13%	99.85%	80.25%	69.37%	90.31%	99.20%	83.33%
16		0.02	77.77%	79.92%	94.46%	99.44%	99.74%	82.67%	70.17%	90.16%	99.25%	83.58%
17		0.03	76.25%	82.18%	93.85%	99.26%	99.69%	83.29%	71.38%	87.94%	98.74%	83.69%
18		0.04	76.52%	81.91%	95.72%	98.61%	99.59%	85.92%	67.81%	88.40%	98.71%	83.03%
19		0.05	77.12%	81.85%	95.01%	98.70%	99.74%	82.76%	68.00%	88.79%	98.61%	82.84%
20		0.06	74.56%	83.11%	94.81%	99.04%	99.74%	85.32%	67.63%	88.21%	98.59%	82.68%

(continued)

Table 7.2 (continued)

Batch#	Err. Tol. (ϵ)	Reg. param (λ)	Classes									Acc (%)
			Corn-No-till	Corn-min	Grass/Pasture	Grass Trees	Hay-Windrowed	Soybeans-notil	Soybeans-min	Soybeans-Clean	Woods	
21	E-05	.006	81.23%	81.50%	91.94%	98.70%	99.95%	76.08%	67.90%	90.39%	99.50%	82.83%
22		.007	77.54%	82.78%	94.01%	98.95%	99.90%	78.99%	69.91%	89.81%	99.41%	83.29%
23		.008	78.16%	80.14%	92.24%	98.86%	100%	78.32%	69.91%	90.82%	99.31%	83.05%
24		.009	78.68%	82.07%	94.51%	99.32%	99.85%	78.87%	69.49%	91.71%	99.25%	83.42%
25		**0.01**	**79.28%**	**82.10%**	**94.01%**	**99.54%**	**99.85%**	**82.17%**	**70.12%**	**90.70%**	**99.30%**	**83.97%**
26		0.02	78.35%	82.48%	95.37%	99.20%	99.79%	81.68%	68.77%	90.12%	99.26%	83.42%
27		0.03	77.78%	83.30%	95.11%	98.79%	99.90%	83.27%	70.01%	88.87%	99.10%	83.78%
28		0.04	76.01%	83.87%	94.66%	98.89%	99.90%	83.23%	70.09%	89.57%	99.13%	83.60%
29		0.05	75.38%	81.96%	96.02%	98.83%	99.54%	82.30%	69.27%	88.33%	98.79%	82.93%
30		0.06	74.93%	81.50%	95.37%	98.76%	99.69%	83.02%	70.95%	87.94%	98.78%	83.31%

Table 7.3 Confusion matrix showing per class accuracy of batch 25

Classes	Corn-No-till	Corn-min	Grass/Pasture	Grass Trees	Hay-Windrowed	Soybeans-notil	Soybeans-min	Soybeans-clean	Woods
Corn No-till	1057	81	3	4	0	64	84	40	1
Corn min	48	604	0	0	0	5	38	39	0
Grass/Pasture	2	2	374	10	2	0	0	6	1
Grass Trees	0	0	2	644	0	0	0	0	1
Hay Windrowed	0	0	0	0	388	0	0	0	1
Soybeans notil	34	34	4	10	0	713	45	28	0
Soybeans min	233	212	12	8	1	171	1660	66	05
Soybeans clean	10	21	1	1	0	7	8	466	0
woods	0	0	5	3	0	0	0	0	1186

resolution of 20 meters is provided) [36]. The accessibility of the ground truth of the HSI dataset, presented in Fig. 7.5a, marks it as one of the typical datasets utilized for classification in the research community. We observe dual situations here in order to evaluate the accuracy. In the initial one, we utilized the complete HSI image for assessment while in the following situation, we utilized a subsection share of the complete image.

7.4.2.1 AVIRIS Full Image

In the initial phase, a complete AVIRIS HSI was taken. That possessed 145 * 145 spatial resolution with 9 distinct classes. The classification performance of the proposed technique can be seen in Table 7.4. It also includes its evaluation with support vector machine, semi-supervised methods, vibrant subspace, neural network, and NS techniques. The randomly chosen training data for every class ranging 3–100 is exploited to be utilized for the vocabulary atoms. This HSI dataset is considered one of the very complex and difficult sets of data in comparison to other datasets for classification purposes as it contains really low spatial resolution. The outcomes are exhibited in Fig. 7.6, while classification maps are demonstrated in Fig. 7.5. One can clearly perceive that the classification performance of the developed methods by means of scarce training data is much improved than NS, neural network, SVM, and semi-supervised methods. If we take 3 labeled data for each class, which implies 27 labeled samples altogether for 9 classes, the classification performance of the proposed general LP scarce and *Homotopy* rooted techniques are 57.67% and 55.10%, correspondingly, in contrast to the precisions of 51.27%, 49.96%, 46.76%, and 54.63% for NS, NN, semi-supervised, and SVM methods, individually. As one can clearly observe, the performance of the proposed technique is much better than the existing associated techniques.

Table 7.4 The results of Semi-Supervised Learning, SVM, Nearest Neighbor, Nearest Subspace, Sparse, and *Homotopy* in AVIRIS Full Image

Labeled Samples	Semi-Supervised	SVM	Nearest Neighbor	Nearest Subspace	Sparse	Homotopy
3	46.76%	54.63%	49.96%	51.27%	**57.67%**	55.10%
5	50.45%	60.49%	54.08%	59.2%	**64.56%**	57.42%
10	52.33%	64.70%	60.80%	65.8%	**69.59%**	67.32%
15	51.05%	68.45%	64.47%	68.46%	**72.11%**	70.62%
20	54.94%	69.30%	67.48%	70.07%	**74.93%**	73.51%
25	54.46%	69.61%	67.81%	69.77%	75.25%	**75.28%**
40	53.47%	71.22%	69.71%	68.72%	75.76%	**76.35%**
100	57.26%	76.21%	75.89%	62.85%	81.07%	**83.71%**

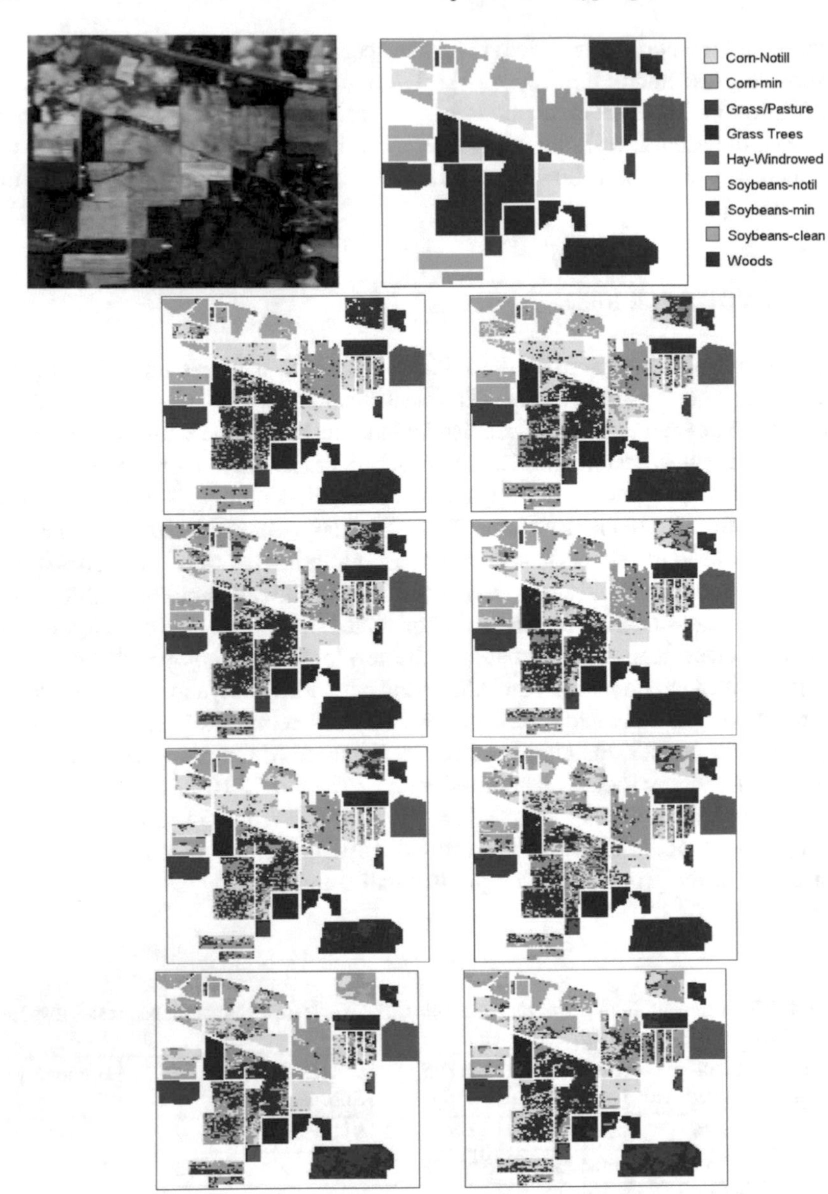

Fig. 7.5 AVIRIS Indian Pines dataset along with the classification maps. **a** Original AVIRIS Indian Pines Image. **b** Ground truth map containing 9 land cover classes. **c**–**j** are the classification maps by using 100, 30, 25, 20, 15, 10, 5, and 3 labeled samples per class

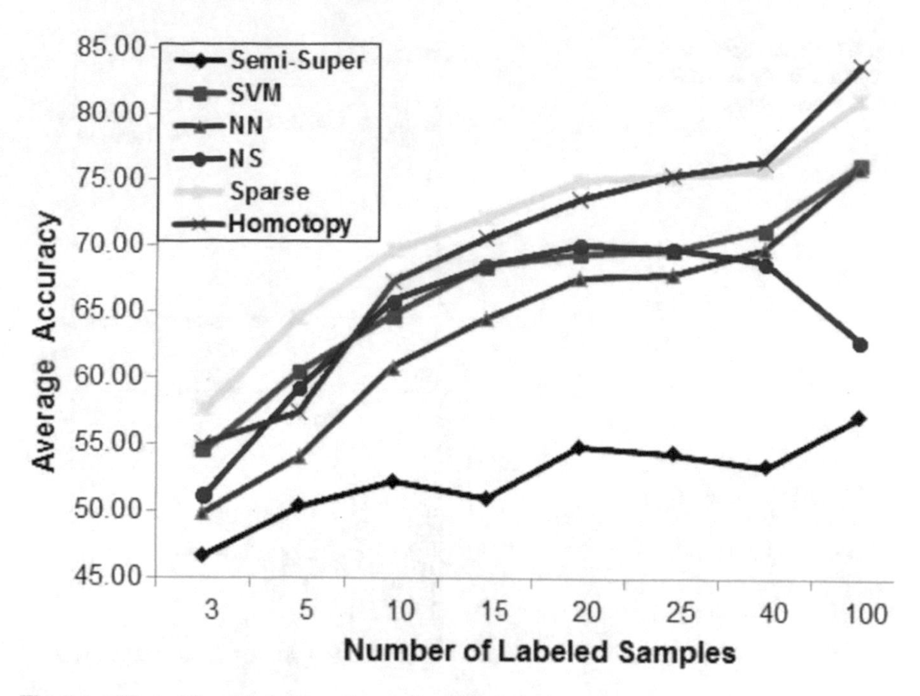

Fig. 7.6 Effect of few labeled samples on classification accuracy

The vector that contains faults η that are formed and explained in Eq. 7.6 is evaluated for every data example . Afterwards, it is deducted in the conclusion phase to perform the classification task for that particular example. The fault matrix which is evaluated for the AVIRIS sensor-based Indian Pines dataset is depicted in Fig. 7.7. One can observe one significant tendency in the accuracies in Table 7.4; the performance for NS techniques are growing when training data for each class is in the limit 3–20. However, if the training data for each class is additionally enlarged in the limit 25–100, the accuracy performance starts to decline instead of having a positive trend. On the other hand, the classification performance of the developed scarce rooted techniques can always experience positive growth. The confusion matrix for 100 examples of training data by utilizing the NS technique and the *Homotopy* method is presented in Fig. 7.8. It can be clearly seen from the matrix that maximum inaccurate performances occur for classes 1, 2, 6, 7, and 8 for NS, and *Homotopy* rooted techniques together. The major problem is that the given data of these mentioned classes have great similarity between each other. Nevertheless, the proportion of wrongly classified examples for the *Homotopy* technique is much less as compared to the wrong classification examples of the NS technique. The major argument is that in the NS technique, a test data x is allocated a class for which the space among x and the subspace covered by the training data of specified class are the least between exclusively the rest of the classes. Since classes 1, 2, 6, 7, and 8 contain a high proportion of similarity between each other, once the training data for each class

Fig. 7.7 Error Vectors **a–d** calculated for the AVIRIS Indian Pines dataset

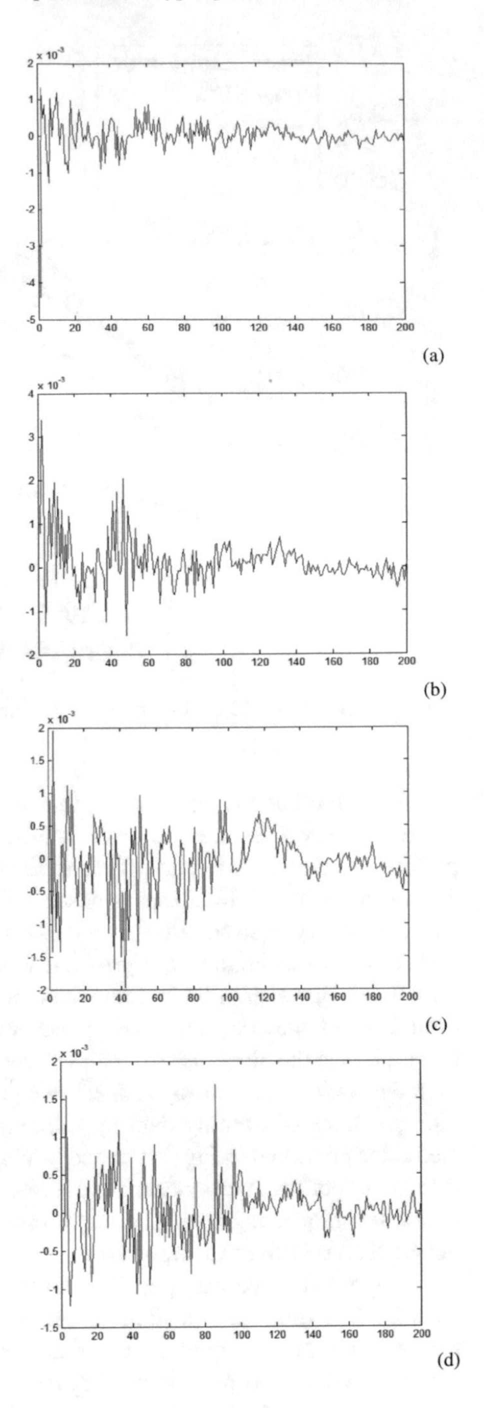

1010	121	1	6	0	60	87	49	0
21	641	0	0	0	3	23	46	0
3	1	369	13	4	0	0	6	1
0	0	1	643	1	0	0	2	0
0	0	0	0	389	0	0	0	0
44	12	1	15	0	717	54	25	0
268	244	8	4	8	191	1599	45	1
1	16	1	4	5	2	14	471	0
0	0	5	5	2	0	0	0	1182

(a)

536	167	3	9	0	237	348	34	0
94	321	4	0	0	53	174	88	0
0	1	357	5	0	7	8	4	15
0	0	30	587	0	8	4	1	17
0	0	2	1	385	0	0	1	0
97	74	4	6	0	459	214	14	0
286	374	24	8	0	349	1193	134	0
17	34	3	0	0	34	99	327	0
0	0	15	6	0	0	0	0	1173

(b)

Fig. 7.8 Confusion matrices of AVIRIS Indian Pines dataset for 100 labeled samples per class using **a** *Homotopy* algorithm **b** NS classifier

is increased, i.e., increase the measurement of subspace, the space among the subspaces develops reduced values. Subsequently, the data examples are not correctly classified. The main logic behind the respectable performance of the ℓ^1-scarce rooted technique in comparison to the NS technique is depicted in [37] which are ℓ^1 rooted techniques are logically biased in comparison to NS-based techniques. Moreover, the said techniques constantly compute the finest and sparest presentation of the training examples in terms of vocabulary atoms.

From the abovementioned performance results, one can observe that *Homotopy* rooted performance is lower than the general LP sparse rooted technique in the presence of much fewer training data. However, if the amount of training data is increased, the results become equivalent to the general LP scarce rooted technique. The technique rooted from *Homotopy* is comparatively greatly robust in comparison to the general LP scarce rooted technique.

7.4.2.2 AVIRIS Sub-Image

In another setup, we get a subsection from the entire AVIRIS image, i.e., widthwise [27–94] and of length [31–116] resulting in a size of 68*86 covering 4 classes. This subsection is also utilized as one of the benchmarks by many researchers for accuracy estimation determination. Table 7.5 displays the classification outcomes of developed methods, which includes generic-determined linear programming resolver and the *Homotopy* rooted process as compared to NS, neural network, support vector machine, and semi-supervised techniques. The training data which is utilized is in the limit of 3–100 for each class. The classification plots are described in Fig. 7.9. As explained in Table 7.5, the overall accuracy of developed algorithms is very useful for limited available training data. For the training data as minimum as 12 examples, it means 3 examples for each class and the generic-determined LP rooted scarce presentation attained 71.59%, while the *Homotopy* rooted scarce presentation attained 69.37%; these results are improved than NS, neural network, support vector machine, and semi-supervised techniques. Taking all the scenarios into account, the precision of the developed techniques is much improved in comparison to the existing associated techniques. As enticipated, if the training data is increased in vacabulary, it greatly effects the classification performance as the accuracy also increases. For a limited training data for each class, that is, 3, 6, 12, etc., the performance of *Homotopy* rooted techniques are marginally smaller as compared to the generic LP scarce rooted technique whereas, as the training data is increased, the performance becomes somewhat similar to the generic LP scarce rooted technique. Though *Homotopy* rooted techniques attained somewhat smaller precision as compared to the BP rooted technique, nevertheless there is an enormous improvement in time, since it monitors the L-step characteristic. Therefore, it is greatly quicker. One can detect a similar performance of the NS technique as once detected in part of a full AVIRIS image. That is, the performance of NS was escalating till 30 training data for each class, however, afterwards the performance begins to decline. This pixel set is a subsection

Table 7.5 The results of SVM, Semi-Supervised, Nearest Neighbor, Nearest Subspace, Sparse, and *Homotopy* in AVIRIS sub-Image

Training samples	SVM	Semi-Super	NN	NS	Sparse	Homotopy
3	58.95%	60.59%	66.95%	65.98%	**71.59%**	69.37%
5	60.63%	62.42%	71.75%	70.87%	**78.28%**	73.65%
10	67.59%	69.15%	75.22%	78.13%	**82.65%**	78.82%
15	74.24%	75.75%	78.77%	80.8%	**84.82%**	82.83%
20	75.35%	76.93%	81.28%	82.3%	**85.88%**	85.61%
25	78.32%	79.85%	80.93%	83.41%	**86.83%**	86.06%
30	78.90%	80.68%	81.10%	83.73%	**87.96%**	86.25%
100	84.50%	84.83%	86.67%	74.79%	90.13%	**90.55%**

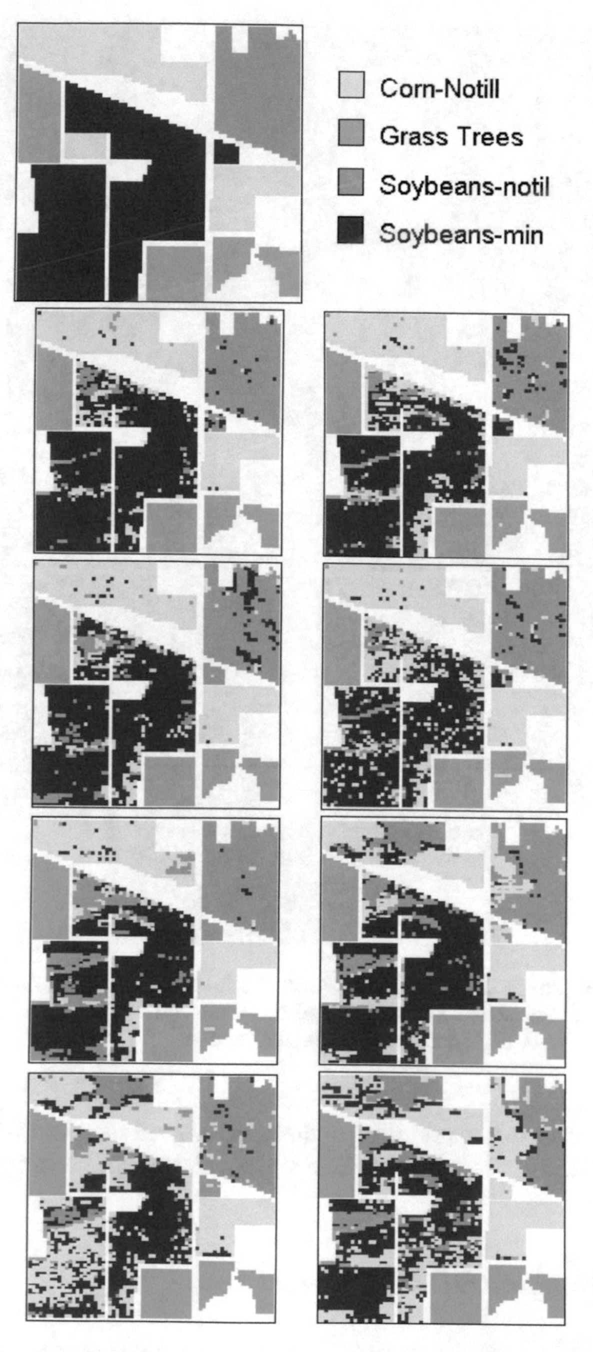

Fig. 7.9 AVIRIS Indian Pines sub-dataset along with the classification maps. **a** Ground truth map containing 4 land cover classes. **b–i** are the classification maps by using 100, 30, 25, 20, 15, 10, 5, and 3 labeled samples per class

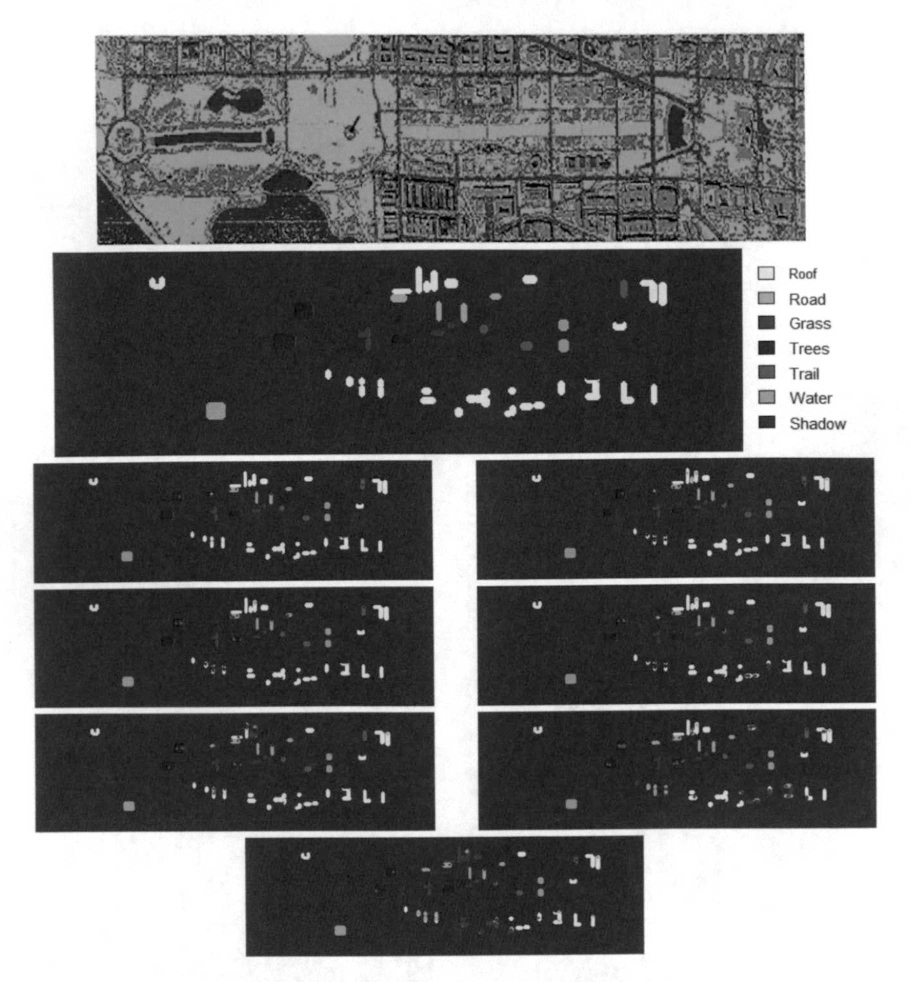

Fig. 7.10 Washington DC Mall dataset along with the classification maps. **a** Original Washington DC Mall Image. **b** Ground truth map containing 7 land cover classes. **c–i** are the classification maps by using 50, 30, 25, 20, 10, 5, and 3 labeled samples per class

of the entire image that was utilized in the preceding subsection. Consequently, the equivalent description is applicable here too, which was designated previously.

7.4.3 Washington DC Mall Image

In this experimental phase, an HSI image of the Washington DC mall is exploited for classification, which is depicted in Fig. 7.10a. This dataset was acquired by utilizing a Hyperspectral Digital Imagery Collection Experiment (HYDICE) sensor.

Table 7.6 The results of SVM, DSMw1, DSMw2, Nearest Neighbor, Nearest Subspace, Sparse, and Homotopy algorithms in WDC hyperspectral image

Labeled samples	SVM	DSMw1 (svm)	DSMw2 (svm)	NN	NS	Sparse	Homotopy
3	75.84%			78.02%	82.25%	79.19%	**84.63%**
5	80.80%			82.94%	84.34%	84.18%	**86.85%**
10	88.05%			94.26%	96.17%	94.27%	**96.41%**
20	92.48%	81.90%	87.80%	96.15%	97.01%	**98.72%**	97.63%
25	93.07%			97.21%	**99.32%**	98.93%	99.13%
40	94.62%	84.60%	92.50%	97.98%	99.31%	**99.49%**	99.27%
50	97.85%			98.43%	99.39%	**99.69%**	99.33%

It comprises 210 spectral channels in the series of 0.4–2.4 um. However, 19 channels were rejected because of water absorption. This hyperspectral image comprises 1280 rows and 307 columns. In the experiment, 7 material classes were utilized that include Rooftop, Road, Trail, Grass, Tree, Water, and Shadow. The subject HSI is comparatively not as problematic as AVIRIS in terms of classification. In order to analyze the accuracy of the developed classification techniques in the presence of limited training, seven multiple extents of training data are utilized for every class in the limits 3–50. The classification plots are depicted in Fig. 7.10. The classification performance of developed techniques including assessments are accessible in Table 7.6. For 3 number of training data for each class, support vector machine, neural network, and NS attained 75.84%, 78.02%, and 82.25%, correspondingly, while the generic LP scarce rooted technique and the *Homotopy* rooted technique attained 79.19% and 84.63% precision, correspondingly. Comparison results for DSM techniques are not accessible for 3, 5, and 10 training data for each class. In an experiment with training data, 20 for each class, the DSM techniques, that is DSMw1 (svm) and DSMw2 (svm), attained 81.90% and 87.80%, correspondingly. Support vector machine attained 92.48%, while the developed scarce rooted technique and *Homotopy* rooted techniques attained 98.72% and 97.63%, correspondingly. Hence, in nearly the entire phases of training data utilized, the developed techniques demonstrated improved performance in the presence of limited training data in comparison to the existing techniques. NS is marginally improved as compared to the developed technique when the training data is increased to 25. Showing tremendous improvement over the existing renowned techniques by utilizing only a limited amount of training data demonstrates that much of the information of HSI is present in low-dimensional subspace. One more thing, one can witness an increase in the performance of the NS technique linearly in association with an increase in training data, in contrast to the situation in AVIRIS datasets. The main reason for this fact is that no similarity is present in each class (which was the case in AVIRIS); even if the dimensionality of the basis is increased, the space among the subspaces is sufficient for the efficient classification.

7.4.4 Kennedy Space Center and Salina A Hyperspectral Datasets

An additional dataset which was utilized for experimentation and evaluation was the HSI of Kennedy Space Center, Florida, as presented in Fig. 7.11a, which was acquired in winter, 1996. It mainly comprises the vegetarian area. The dataset consists of the spectral resolution of 224 spectral channels. The entire amount of spectral channels utilized were around 176 after removing the 48 bands full of noise due to water absorption. There were 13 different classes in total in the respective HSI image and there were approximately 5211 pixels of labeled training samples altogether for 13 classes. The classification performance is described in Table 7.7, while classification plots are depicted in Fig. 7.11. Very limited training data in the span of 3–50 are exploited for inspecting the accuracy of the developed techniques. By utilizing 3–5 training examples for each class, support vector machine attained 61.09% and 73.02% correspondingly, the generic LP scarce rooted technique attained 77.47% and 81.22%, correspondingly, and on the other hand, the *Homotopy* rooted technique attained 77.91% and 81.08%, individually. In all the different experimentations, developed techniques performed much better than the existing techniques in comparison.

A hyperspectral image of Salina A was acquired by Airborne Visible Infrared Sensor (AVIRIS) over the valley of Salinas, California. It consists of 224 channels. It contains a great spatial resolution of 3.7 meters. It includes 86 rows by 83 columns. The HSI image is presented in Fig. 7.12a. Similar to Indian Pines and other AVIRIS sensor-based images, around 20 channels full of noise need to be removed. In this image, these channels include [108–112], [154–167], and 224. This HSI image consists of multiple classes including vegetables, bare soils, vineyard fields, etc. Classification plots are described in Fig. 7.12, while classification performance is depicted in Table 7.8. In this particular image, all the techniques performed really well and the precision of all the techniques is very high.

Table 7.7 The results of SVM, Nearest Neighbor, Nearest Subspace, Sparse, and *Homotopy* algorithms in Kennedy Space Center hyperspectral image

Labeled samples	SVM	Nearest Neighbor	Nearest Subspace	Sparse	Homotopy
3	61.09%	70.72%	71.52%	77.47%	**77.91%**
5	73.02%	76.59%	78.49%	**81.22%**	81.08%
10	80.83%	80.48%	84.53%	84.98%	**85.14%**
20	83.18%	84.02%	86.53%	**87.52%**	87.36%
25	85.20%	84.48%	87.41%	87.88%	**88.24%**
30	86.09%	85.22%	87.48%	88.80%	**89.25%**
50	87.55%	87.03%	87.63%	90.46%	**90.96%**

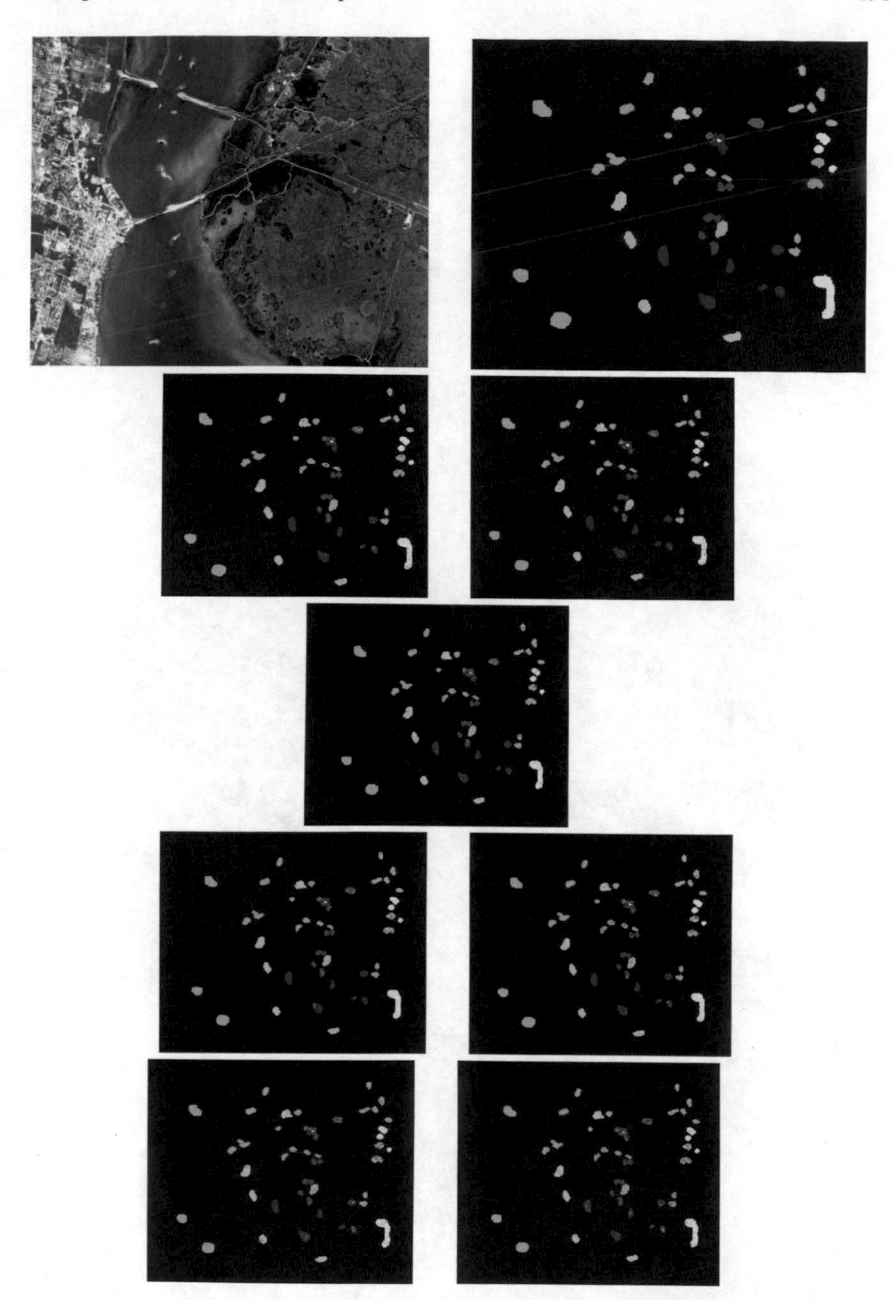

Fig. 7.11 Kennedy Space Center dataset along with the classification maps. **a** Original Kennedy Space Center Image. **b** Ground truth map containing 13 land cover classes. **c**–**i** are the classification maps by using 50, 30, 25, 20, 10, 5, and 3 labeled samples per class

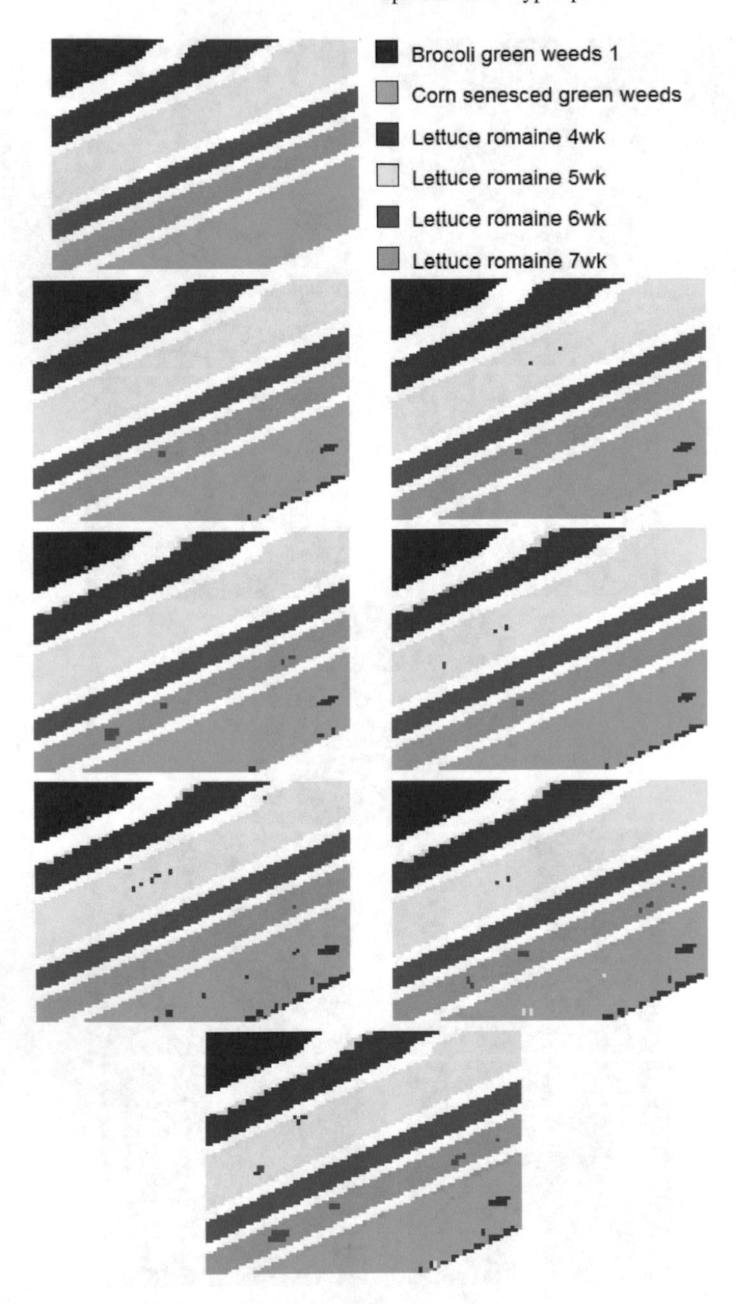

Fig. 7.12 Salina A dataset along with the classification maps. **a** Original Salina Image. **b** Ground truth map containing 7 land cover classes. **c**–**i** are the classification maps by using 50, 30, 25, 20, 10, 5, and 3 labeled samples per class

Table 7.8 The results of SVM, Nearest Neighbor, Nearest Subspace, Sparse, and *Homotopy* algorithms in Salina hyperspectral image

Labeled samples	SVM	Nearest Neighbor	Nearest Subspace	Sparse	Homotopy
3	83.75%	91.09%	93.53%	95.92%	**95.96%**
5	90.11%	94.91%	95.86%	**98.06%**	97.49%
10	97.20%	97.47%	98.58%	**98.77%**	98.41%
20	98.85%	98.28%	**99.21%**	98.89%	98.66%
25	99.19%	98.56%	**99.23%**	99.19%	99.09%
30	99.36%	98.68%	**99.42%**	99.23%	99.40%
50	99.45%	98.95%	99.55%	99.36%	**99.58%**

7.4.5 *Time Comparison of General LP Sparse and* Homotopy*-Based Sparse Representations*

In this sub-part of the chapter, the time required to perform the classification task of datasets including AVIRIS sensor-based sub-part of Indiana Pine shown in Fig. 7.9 and AVIRIS Indian Pine full HSI as shown in Fig. 7.8 is correlated. These images are evaluated from a time perspective by utilizing generic-aimed LP resolver and *Homotopy* rooted scarce presentations. For the final results, the first ten experiments are conducted and the mean of these ten experiments is taken as the final result. Time consumption of the AVIRIS sensor-based dataset is to classify the test data in comparison to the training data for each class in the vocabulary as shown in Figs. 7.13 and 7.14. One can clearly observe from the figure the time required to classify the test examples, of generic LP resolvers, rises rapidly and aggressively with an increase in the labeled training data. Consequently, it is easy to conclude that the *Homotopy* rooted technique is very robust in terms of time in comparison to the conventional generic LP resolvers.

7.5 Discussion

In this section, we discuss various aspects of sparse-based approximation using l_1-*minimization*.

7.5.1 *Sparsity of Computed Solution*

The scarce rooted classification technique is beneficial if the data provided is scarce in nature. In this technique, the testing data is estimated by linearly combining the

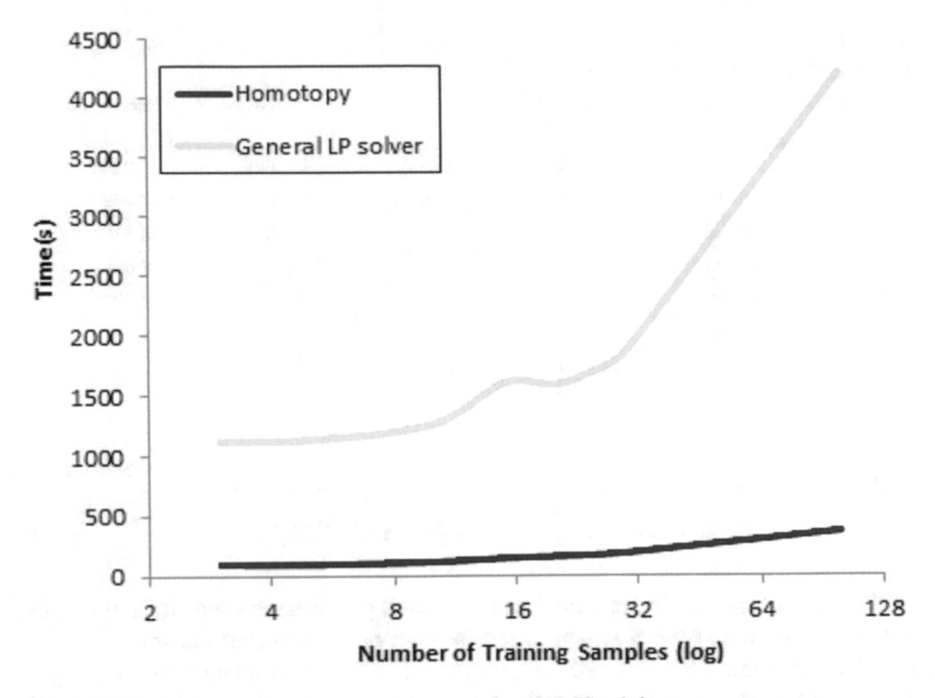

Fig. 7.13 Time comparison of proposed approach for AVIRIS sub-image

dictionary items. The parameters that are generated during this process are utilized afterwards for restoration and classification goals. The scarcer the test data is, the scarcer is their estimation. Figure 7.15 represents the performance of the *Homotopy* technique for AVIRIS- based Indian Pine HSI in order to demonstrate the scarcity of the HSI data. The labeled training data is in the range of 3–100 data for each class. In Fig. 7.15a, 100 labeled training examples for each class were utilized, while the *Homotopy* technique was utilized to compute the scarce presentation, i.e., parameter vector α. In Fig. 7.15b, the rebuilding leftover are presented for diverse classes. Observe the scarcity of the resolution computed by the *Homotopy* technique in Fig. 7.15a. Only very few scarce vector elements are not 0. This proves that fewer vector parameters in the sample for each class were essential for the presentation of test examples by the *Homotopy* technique.

The utmost significant constraint for the *Homotopy* technique is to function with the high scarce resolution of the test data [4, 24]. The *Homotopy* technique possesses an L-step resolution characteristic. That means if the fundamental resolution is extremely scarce, and comprises merely L non-zeros, the technique accomplishes that scarcest resolution in merely L repetitive stages [24]. Once this characteristic exists, and L is extremely minor, this implies that ℓ^1-minimization complications having L-scarce resolutions can be resolved in a segment of the price of resolving the entire linear scheme [24]. From the decent classification performance of the

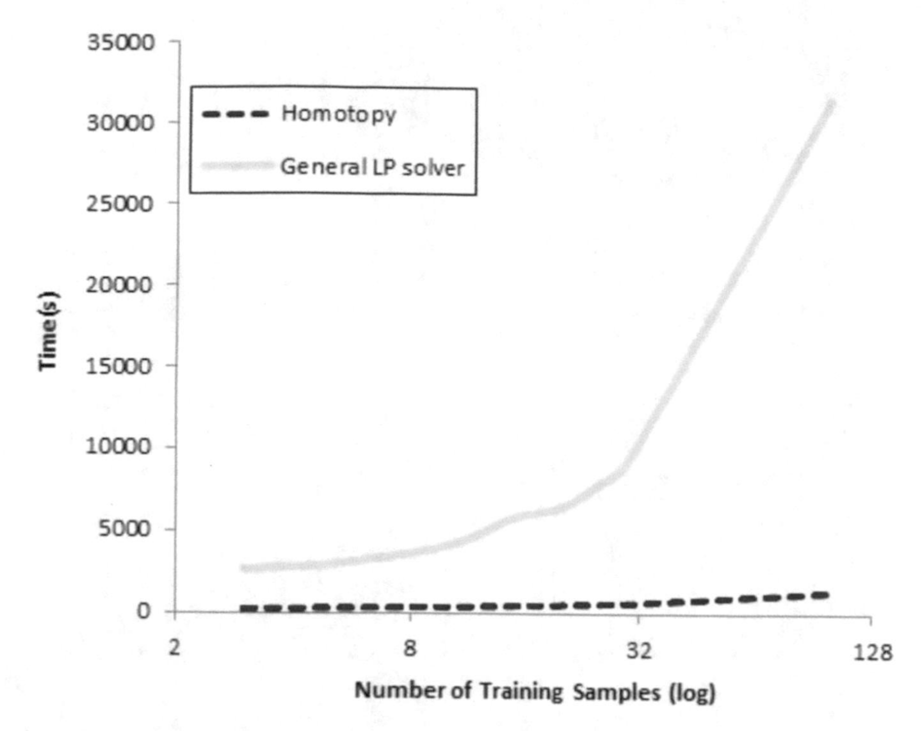

Fig. 7.14 Time comparison of proposed approach for AVIRIS full image

Homotopy technique, it can easily be concluded that HSI spatial pixels are extremely scarce in nature.

It is well established among the researchers that the major reason for any algorithm to work in an extremely great dimensionality scenario is only because in reality, the given data is not actually high-dimensional [38]. Somewhat fairly, it deceits in a place of a fairly lesser dimensionality [38]. Principal Component Analysis is one of the extremely famous techniques for the inherent dimensional approximation of data [39]. Experiments also shown in Fig. 7.4 demonstrate that an HSI image deceits in a lower dimensionality subspace. Furthermore, an Eigenvector rooted inherent dimensionality approximation technique [39] is exploited to approximate the dimensional of HSI datasets, and the outcomes of inherent dimensionality for AVIRIS, Washington DC, Kennedy Space Center, and Salina A datasets are 3, 2, 3, and 4, correspondingly. The outcomes of these techniques demonstrate our supposition that HSI image data deceits in a lower dimensionality. Furthermore, comparable outcomes of lower dimensionality subspace for HSI data have also been demonstrated in [40].

Extremely scarce resolution of HSI data discloses additional components about the properties of HSI pixels that precisely limited characterized training data is sufficient to acquire the variation of testing data. One can assume that information contained in every class is precisely analogous, that is due to the characteristics of

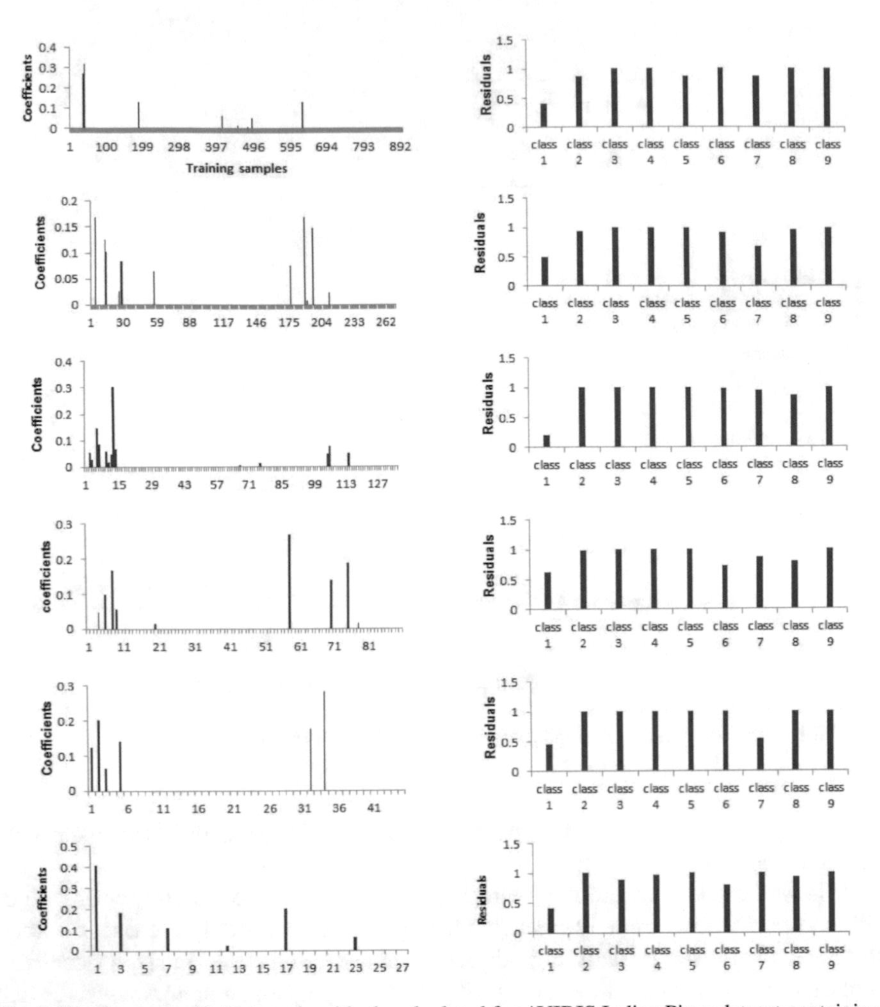

Fig. 7.15 Sparse coefficients and residuals calculated for AVIRIS Indian Pines dataset containing 9 land cover classes using *Homotopy* algorithm **a** Sparse coefficients for 100 training samples per class with average accuracy 83.71%. **b** Residuals for 100 training samples per class for 9 classes. **c**, **d–k**, and **l** are sparse coefficients and residuals of 9 classes for 30, 15, 10, 5, and 3 labeled samples per class with average accuarcy 76.35%, 70.62%, 67.32%, 57.42%, and 55.10%, respectively

HSI information as every class presents a single category of substance present on the exterior. Consequently, they should have related reflective signatures. Thus, it is highly likely that testing data of a class is extremely analogous to the labeled training data of that particular class. Hence, extremely limited demonstrative examples are going to be sufficient to accommodate the variation in the testing examples.

Excellent classification outcomes that developed scarce rooted methods demonstrate that the supposition of our approach about lower dimensional lined subspace is accurate. With this supposition, our scarce rooted technique can produce the scarcest resolution by utilizing extremely limited items from the dictionary.

Henceforth, the description for the excellent outcomes of the developed technique is that due to lower dimensionality lined subspace supposition, the effective information of each class falls in a lower dimensional. Hence, the amount of foundation vectors required, that is the sum of non-zero entrances required, is comparable to the dimensionality of the subspace of the class. It implies that if the data follows the low linear subspace assumption then the solution should be highly sparse. Figure 7.4 shows the drive enclosed by 5 Principal Components of each class. PCA of AVIRIS Indian Pines HSI is also delivered, which clearly explains that extremely limited PCA are sufficient to accomplish the major part of energy. All the evidence demonstrated above verify the extremely lower dimensionality space assumption for HSI images.

7.5.2 Sparse Classification

Figure 7.16 establishes a performance of classifying test HSI pixels. The test examples were taken from the Washington DC Mall HSI image. First, fifty training data examples from the entire seven classes of HSI were stacked together forming a matrix A. Afterwards, scarce resolution by exploiting l_1*-minimization* is computed. The parameters in the scarce resolution matching the training data examples of 7 classes are then utilized to restructure the real testing example. Henceforth, we contain 7 estimations. One can observe this in Fig. 7.16 that the nearest estimations are built by utilizing the training examples of the rooftop class with the associated parameters in the scarce resolution matrix, as the test examples are associated with the rooftop class. The rest of the estimations are not comparable. The reason behind not being comparable is that these are built by utilizing the training examples from dissimilar classes as compared to the real testing example class.

7.5.3 Sparse Solution Analysis

This section of the chapter focuses on discussing the issues, challenges, and difficulties faced during the classification of HSI images. The classification results to be discussed here are categorized as good, best and bad scarce HSI results. Figure 7.17 and Table 7.9 demonstrate 3 scenarios of scarce presentation rooted classification for HSI pixels. Coming toward the first scenario, Fig. 7.17a and first row of Table 7.9, the scarce presentation of a testing HSI pixel that is computed by utilizing *Homotopy algorithm* is an improved version because even though the parameters in the scarce resolution are present for numerous classes, nevertheless, the parameters are more than the other for one class. The leftover for class one that is computed by utilizing Eq. (7.8), as can be observed in the first row of Table 7.9, is evidently lesser as compared to the rest of the classes, hence one can allocate the testing pixel to class

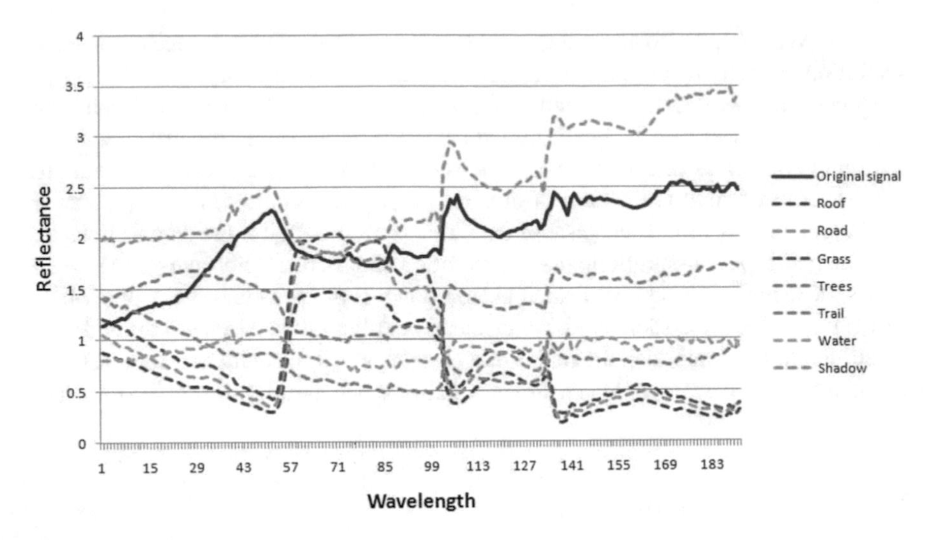

Fig. 7.16 Classification of a hyperspectral test pixel using l_1-*minimization*

Table 7.9 Sparse analysis

Class	1	2	3	4	5	6	7	8
Residual(a)	0.23	0.92	1.0	1.0	1.0	1.0	0.90	0.94
Residual(b)	1.0	1.0	1.0	0.01	1.0	1.0	1.0	1.0
Residual(c)	0.81	0.695	1.0	1.0	1.0	0.697	1.0	0.80

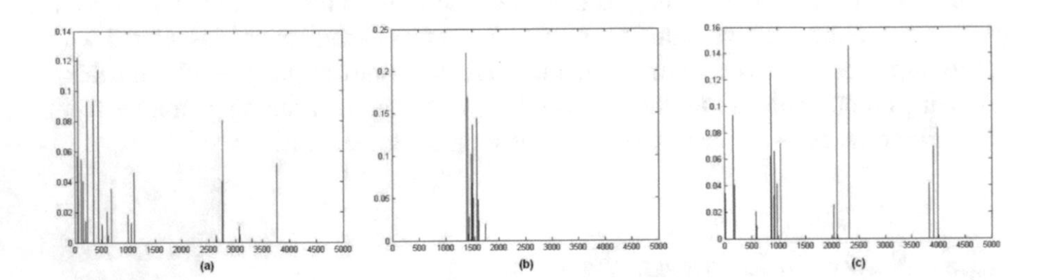

Fig. 7.17 A demonstration of **a** Good **b** Best **c** Bad hyperspectral classification sparse solutions

one. Discussing the second scenario, Fig. 7.17b and second row of Table 7.9, the scarce presentation of HSI testing pixel is excellent and can be considered as one of the perfect scenarios. Here, the parameters in the scarce resolution pertain merely for *A* class, and that class is class four. The leftover for particular class 4, as can be observed in the second row of Table 7.9, is evidently lowermost in comparison to the rest of the classes. Hence, one can straightforwardly allocate the testing pixel to class four.

Discussing the third and last scenario, Fig. 7.17c and the third row of Table 7.9, the scarce presentation of HSI testing pixel is the poorest situation. Here, the parameters

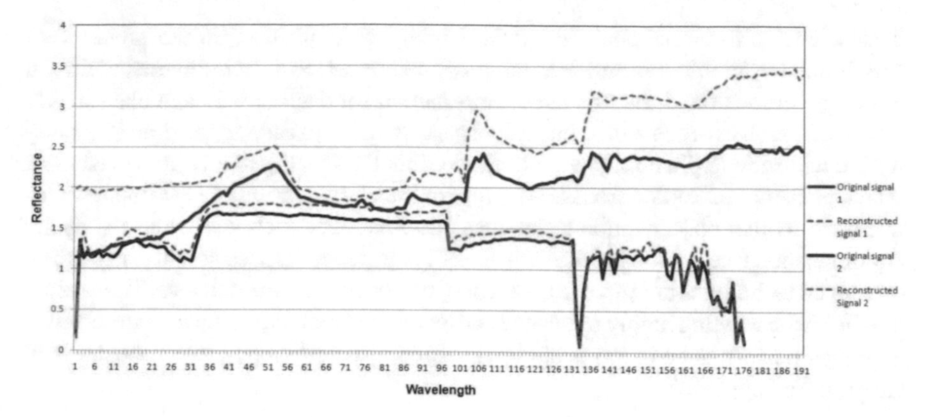

Fig. 7.18 Reconstruction of two hyperspectral test pixels using l_1*-minimization* for 100 training samples

in the scarce resolution not merely pertaining to numerous classes, nonetheless, are also nearly similar for numerous classes. The leftovers are also fairly alike for numerous classes, because one can clearly observe in the third row of Table 7.9. Subsequently, if the measure of the lowermost leftover is assigned here, formerly class two would be given to this particular HSI pixel. That is going to be incorrect because the accurate class assignment is class six. Henceforth, in this scenario the scarce doesn't work. This problematic scenario is the reason for many causes but the major reason behind such a difficult situation is the noise.

7.5.4 *Sparse Reconstruction*

Figure 7.18 presents the restoration of 2 HSI testing examples from the Washington DC Mall HSI image including their finest estimation computed by utilizing l_1-*minimization*. The restoration is achieved by utilizing a hundred training examples for each class. The hard stroke is the real pixel and the dotted strokes are the estimations by utilizing l_1-*minimization*. It is apparent from the figure that l_1-*minimization* proficiently computes the estimation of the HSI testing examples.

7.6 Summary

This chapter of the book explores the comprehensive research efforts on ℓ^1-***minimization*** rooted scarce presentation methodology for the effective classification hyperspectral images by utilizing a limited high dimensionality labeled examples. By manipulating the special characteristics of HSI information, the technique incapacitates a foremost challenge in HSI image classification, in the presence of very limited available training data samples and the curse of high dimensionality. It has been demonstrated that the major information for each HSI image class nearly presents in a lower dimensional linear subspace. In order to cater to the time criticality for the real-time applicability, the effort additionally developed an enormously fast scarce presentation rooted from *Homotopy* for the classification of HSI images. Different from other classification techniques, the developed technique doesn't require any dimensional reduction process and doesn't require model assortment criteria. The developed techniques are verified on 4 real hyperspectral image datasets. The evaluations with the existing highly renowned and developed techniques have demonstrated improved performance in terms of classification, time efficiency, and robustness of the proposed approaches.

References

1. Landgrebe DA (2003) Signal theory methods in multispectral remote sensing. Wiley, Hoboken, NJ
2. Lu D, Weng Q (2007) A survey of image classification methods and techniques for improving classification performance. Int J Remote Sens 28(5):823–870
3. Civco DL (1993) Artificial neural networks for land-cover classification and mapping. Int J Geophys Inf Syst 7(2):173–186
4. Wright J, Yang A, Ganesh A, Sastry S, Ma Y (2009) Robust face recognition via sparse representation. IEEE Trans Pattern Mach Intell 31(2):210–227
5. Chen Y, Nasrabadi NM, Tran TD (2010) Classification for hyperspectral imagery based on sparse representation. In: IEEE proceedings of the WHISPERS, vol 2, pp 1–4
6. ul Haq QS, Shi L, Tao L, Yang S (2010) Hyperspectral data classification via sparse representation in homotopy. In: Proceedings of IEEE international conference on information science and engineering, pp 3748–3752
7. Gualtieri JA, Chettri SR, Cromp RF, Johnson LF (1999) Support vector machine classifiers as applied to aviris data. In: Proceedings of the 8th JPL airborne geoscience workshop, pp 221–232, Feb 1999
8. Scholkopf B, Smola A (2002) Learning with kernels support vector machines, regularization, optimization and beyond. MIT Press Series, Cambridge, MA
9. Yang J-M, Kuo B-C, Yu P-T, Chuang C-H (2010) A dynamic subspace method for hyperspectral image classification. IEEE Trans Geosci Remote Sens 48(7):2840–2853
10. Cand'es EJ, Wakin M (2008) An introduction to compressive sampling. IEEE Signal Process Mag 25(2):21–30
11. JingYu Y, YiGang P, WenLi X, QiongHai D (2009) Ways to sparse representation: an overview. Sci China Ser F-Inf Sci 52(4):695–703
12. Donoho DL (2006) For most large underdetermined systems of linear equations the minimal. L1-norm solution is also the sparsest solution. Comm Pure Appl Math 59(6):797–829
13. Naseem I, Togneri R, Bennamoun M (2009) Sparse representation for video-based face recognition. Springer, Berlin, Heidelberg, pp 219–228
14. Gemmeke JF, Cranen B (2008) Noise reduction through compressed sensing. In: Proceedings of Interspeech, pp 1785–1788
15. Huang K, Aviyente S (2006) Sparse representation for signal classification. Neural Inf Process Syst
16. Candes EJ, Romberg J, Tao T (2006) Robust uncertainty principles: exact signal reconstruction from highly incomplete frequency information. IEEE Trans Inf Theory 52(2):489–509

17. Candes EJ, Tao T (2006) Near-optimal signal recovery from random projections: universal encoding strategies? IEEE Trans Inf Theory 52(12):5406–5425
18. Donoho DL (2006) Compressed sensing. IEEE Trans Inf Theory 52(4):1289–1306
19. Tibshirani R (1996) Regression shrinkage and selection via the lasso. J R Stat Soc Ser B 58(1):267–288
20. Chen Y, Nasrabadi NM, Tran TD (2010) Sparse subspace target detection for hyperspectral imagery. In: SPIE proceedings of the algorithms and technologies for multispectral, hyperspectral and ultrspectral, p 3
21. Chen Y, Nasrabadi NM, Tran TD (2010) Sparsity-based classification of hyperspectral imagery. In: IEEE proceedings of the IGARSS, pp 2796–2799
22. Castrodad A, Xing Z, Greer J, Bosch E, Carin L, Sapiro G (2010) Discriminative sparse representations in hyperspectral imagery. In: 2010 17th IEEE international conference on image processing (ICIP), pp 1313 –1316, Sept 2010
23. Malioutov DM, Cetin M, Willsky AS (2005) Homotopy continuation for sparse signal representation. In: Proceedings of the IEEE international conference on acoustics, speech, and signal processing, pp 733–736
24. Donoho D, Tsaig Y (2006) Fast solution of l1-norm minimization problems when the solution may be sparse. Tech. rep., Department of Statistics, Stanford University
25. Plumbley MD (2005) Geometry and homotopy for l1 sparse representations. In: Proceedings of SPARS'05, Rennes, France, pp 206–213, Nov 2005
26. Majumdar A, Ward RK. Nearest subspace classifier: application to character recognition. [Online] http://ubc.academia.edu/documents/0009/0488/NSC.pdf
27. Elad M, Aharon M (2006) Image denoising via sparse and redundant representations over learned dictionaries. IEEE Trans Image Process 15(12):3736–3745
28. Camps-Valls G, Bandos T, Zhou D (2007) Semi-supervised graph-based hyperspectral image classification. IEEE Trans Geosci Remote Sens 45(10):2044–3054
29. Amaldi E, Kann V (1998) On the approximability of minimizing nonzero variables or unsatisfied relations in linear systems. Theor Comput Sci 1:209:237–260
30. Figueiredo MAT, Nowak RD, Wright SJ (2007) Gradient projection for sparse reconstruction: application to compressed sensing and other inverse problem. IEEE J Sel Top Signal 1:586–597
31. Candes E, Romberg J (2006) l1–magic: a collection of matlab routines for solving the convex optimization programs central to compressive sampling
32. Tropp JA, Gilbert AC (2007) Signal recovery from random measurements via orthogonal matching pursuit. IEEE Trans Inf Theory 53(12):4655–4666
33. Yang J, Zhang Y (2009) Alternating direction algorithms for ℓ_1-problems in compressive sensing. No. TR09-37
34. Donoho DL, Tsaig Y (2006) Fast solution of ℓ_1-norm minimization problems when the solution may be sparse. Tech. Rep. Stanford CA, 94305
35. Osborne MR, Presnell B, Turlach BA (2000) A new approach to variable selection in least squares problems. IMA J Numer Anal 20(11):389–403
36. Plaza A, Benediktsson JA, Boardman JW, Brazile J, Bruzzone L, Camps-Valls G, Chanussot J, Fauvel M, Gamba P, Gualtieri A, Marconcini M, Tilton JC, Trianni G (2009) Recent advances in techniques for hyperspectral image processing. Remote sensing of environment, vol 113, pp 110–122, Sept 2009
37. Yang AY, Wright J, Ma Y, Sastry S (2007) Feature selection in face recognition: a sparse representation perspective. Tech. Rep. UCB/EECS-2007-99
38. Levina E, Bickel PJ (2004) Maximum likelihood estimation of intrinsic dimension. In: NIPS
39. van der Maaten L, Postma EO, van den Herik HJ (2008) Dimensionality reduction: a comparative review
40. Bioucas-Dias JM, Nascimento JMP (2008) Hyperspectral subspace identification. IEEE Trans Geosci Remote Sens 46(8):2435–2445

Chapter 8
Challenges and Future Prospects

The demand for accurate and robust algorithms for the analysis of hyperspectral remote sensing images has increased significantly due to the recent advancements in remote sensing technology. The hyperspectral image classification involves target detection of different ground covers on the surface of the earth and the categorization of the subject's geographical area into different classes of interest. The classification of a hyperspectral remote sensing scene is a challenging task due to various reasons. First of all, it is a very complex procedure that involves different processes aiming at extracting and analyzing all the rich spectral and spatial materials enclosed in the hyperspectral image. Secondly, the very complex data is the integration of spectral and spatial information with the Hughes phenomena, very limited labeled samples, and redundancy with inherent sensor and environmental noise.

By keeping the above challenges in view, this book focuses on developing techniques for analyzing the detailed available spectral and spatial information, extracting the discriminative and invariant spectral and spatial features. A hyperspectral image comprises hundreds of spectral bands with a significant amount of redundancy and noise that not only makes the subsequent analysis of HSI really challenging and difficult but also degrades the performance of subsequent classification algorithms. Therefore, the first step toward classification is performed (Chap. 3) by analyzing the spectral channels in detail. We detect the noise and redundancy in all the bands by adaptive boundary movement criteria, which results in identifying the inherent noise and redundancy present in the data. Then, each band is classified into different categories in terms of the spectral information contained in it. This is an essential step in order to get a profound insight into the data and meet the challenges of high dimensionality, redundancy, and noise. The removal of redundancy and noise is a vital step for all subsequent applications such as segmentation, feature selection, target detection, classification, and unmixing. The results proved that the identifica-

L. Tao and A. Mughees, *Deep Learning for Hyperspectral Image Analysis and Classification*, Engineering Applications of Computational Methods 5,
https://doi.org/10.1007/978-981-33-4420-4_8

tion and removal of noise and redundancy significantly improve the performance of subsequent classification algorithms.

Ultimately, we developed accurate deep learning-based HSI classification techniques for land cover classes. After exploiting the spectral information that hyperspectral images deliver, the research was directed toward the exploitation of spatial information (Chap. 4). In recent years, researchers have focused on the importance of incorporating spatial information along with spectral data for improved performance in the HSI classification process as recent hyperspectral sensors can deliver excellent spatial resolution along with the spectral channels. A novel strategy for extracting spatial information based on an unsupervised adaptive boundary adjustment-based technique which groups the spatially similar regions through the adaptive boundary adjustment-based criteria was developed. The algorithm was built on a clustering-based method that adaptively adjusts the boundary, resulting in the extraction of geometrical features that relate to expressive structures in the scene.

In the next phase of the research (Chap. 5), the aforementioned findings were exploited for developing a methodology which takes maximum advantage of the extracted spectral and spatial features, merging both spectral and spatial features into a deep learning (DL) framework. The most discriminative and invariant features, extracted by the feature selection technique along with the meaningful geometrical structures, identified by the spatial feature extraction technique, were processed and exploited through different deep learning architectures for improved HSI classification accuracy even with a few available samples. It is worth mentioning that by fusing spectral and spatial features and adopting the adaptive window size instead of the fixed window size in DL, algorithms achieved higher classification accuracies by minimizing the effect of the abovementioned challenges. Three different deep learning architectures were explored and developed by integrating the spatial–contextual features before the classification process for improved HSI classification performance.

In the next phase of the book (Chap. 6), the complexity of the hyperspectral bands is addressed by spectral segmentation, where spectrally similar contiguous bands are grouped together and the DL-based architecture is applied to each spatial–spectral segmented group of bands, separately. Locally applying DL-based feature extraction to each group of bands reduces the computational complexity and simultaneously results in better features, and hence improved classification accuracy is obtained. In the second part of the book, the integration of spatial information is addressed by investigating the methods based on merging the spatial information after the pixel-wise classification through a majority voting-based decision rule. The resulting extracted spatial information through phase 2 (Chap. 4) and DL-based pixel-wise classification results are merged together through majority voting to provide the final classification results. In developed methods, a labeled class for each pixel is decided on the classification result for the target cell, and spatial data is obtained through segmentation.

In Chap. 7, we turn our focus toward the unsupervised classification based on the sparse characteristics of the HSI data. We believe it is good to see both supervised and unsupervised approaches desires of having a classifier which can work in high-dimensional space, and for a few labeled samples. In this chapter, unlike traditional supervised methods, the proposed unsupervised classification method does not have separate training and testing phases and, therefore, does not need a training procedure for model creation. We exploit certain special properties of hyperspectral data and propose an ℓ^1-***minimization***-based sparse representation classification approach to overcome the availability of the limited labeled training data problem in HSI. We assume that the data within each hyperspectral data class lies in a very low-dimensional subspace, and prove the sparsity of hyperspectral data. To handle the computational intensiveness and time demand of general-purpose LP solvers, we propose a *Homotopy*-based sparse classification approach, which works efficiently when data is highly sparse. The approach is not only time efficient, but it also produces results, which are comparable to the traditional methods.

8.1 Future Prospects

HSI analysis is a hot area in remote sensing data analysis due to the vast source of information contained within images. This helps a great deal in understanding and analyzing the properties and characteristics of the earth's surface by integrating rich spectral and spatial information. However, HSI poses major challenges for supervised classification methods due to the high dimensionality of spectral channels and limited available training samples. Researchers have testified some distinctive geometrical, statistical, and asymptotical characteristics of high-dimensional information with the aid of experimental data such as with increase in dimensionality, the volume of a hypercube focuses in corners. Uncertainties caused at diverse phases of information acquirement and the classification process can intensely affect the classification performance. Similarly, it has also been proven that spatial resolution greatly affects the classification accuracies. The incorporation of effective spatial information not only detects the small objects effectively but also reduces the effect of mixed pixels and noise.

This book presents innovative methodologies for the analysis of hyperspectral images by addressing the abovementioned challenges. The experiments performed during the research identified a series of potential enhancements that are encouraging directions for future research work.

- The proposed band categorization/noise removal strategy can be improved in a number of ways. First, the number of clusters which are initialized in the start is selected manually based on the level of complexity in the image. An automatic cluster selection criteria could be devised to automatically select the number of clusters for the HSI scene. This enhancement would enable us to acquire a fully automatic and parameter-free methodology. Moreover, the noise detection strategy is based on three factors, i.e.; cluster-size factor, cluster-shift factor, and cluster spatial–spectral contextual difference factor; more factors can be taken into account in order to improve the accuracy of the technique.

- In the spatial information extraction phase, the size of each cluster is adjusted manually based on the size of the smallest structure in the image. If the size of each cluster is adjusted automatically, small spatial structures are often not identified as distinct areas. This results in the merging of these small regions with the larger neighboring structures. Hence, different methods can be explored in order to enhance this area of the proposed spatial extraction technique. The size of the cluster should be selected automatically such that it incorporates the smallest spatial structure present in the scene.
- In the classification phase, spatial information is merged with the spectral characteristics in a vector. Different merging criteria can be explored to effectively merge the spatial and spectral features. Moreover, a thorough research could also be done on merging the spectral and spatial features during the classification/feature extraction phase into the feature extractor for more effective feature extraction to improve the stability of the classification accuracy.
- In recent years, sparse representation-based classification (SRC) methods have also contributed to dealing with small training sample problems in hyperspectral image classification. The objective of SRC is to represent an unknown sample precisely consuming only a tiny number of atoms in a dictionary. The class label with the minimal representative error will be assigned to the unlabeled samples. Sparse coefficients acquired through SR-based modeling of the hyperspectral images hold discriminative characteristics that can be exploited for hyperspectral image processing tasks, such as classifications. Furthermore, such discriminative features can be significantly improved by incorporating the contextual–spatial information in the sparse coefficient domain. To further explore contextual information and conform the spatial structure as far as possible, the shape-adaptive algorithm developed and presented in this book can be effectively utilized to construct a shape-adaptive region for each test pixel.

Hyperspectral remote sensing technology is advancing significantly in the current time. Sensors, both onboard airborne and space-borne platforms, cover large areas of the earth's surface with unprecedented spectral, spatial, and temporal resolutions. The advancement and availability of HSI sensors are making access to these sensors for commercial use. These characteristics enable a myriad of applications requiring the fine identification of materials or estimation of physical parameters. Very often, these applications rely on sophisticated and complex data analysis techniques. How to extract valuable characteristics from the data. Thanks to Deep Learning theory that is equipped with the capability of automatically extracting and learning the fruitful features from the training set; unsupervised feature learning from very large raw-image datasets has become possible. Actually, DL has proven to be a new and exciting tool that could be the next trend in the development of HSI image processing. Despite its great potential, DL cannot be directly applied in HSI analyses due to the complex, large number of channels and other statistical and geometrical uncertainties. It contains hundreds of bands that can cause a small patch to be a really large data cube, which corresponds to a large number of neurons in a pre-trained network. In addition to the visual geometrical patterns within each band, the spectral curve vectors across bands are also important information. However, how to utilize this information still requires further research. Problems still exist in increasingly high spatial resolution as to how to meaningfully extract the useful information. Furthermore, images acquired by different sensors present large differences. How to transfer the pre-trained network to other images is still unknown.

Printed in the United States
by Baker & Taylor Publisher Services